Student Workbook to Accompany

Power Equipment Engine Technology

Michael Ross

 Cengage

Australia • Brazil • Canada • Mexico • Singapore • United Kingdom • United States

Student Workbook to Accompany Power Equipment Engine Technology

Michael Ross

Vice President, Career and Professional Editorial: Dave Garza

Director of Learning Solutions: Sandy Clark

Executive Editor: David Boelio

Managing Editor: Larry Main

Senior Product Manager: Matthew Thouin

Editorial Assistant: Jillian Borden

Vice President, Career and Professional Marketing: Jennifer McAvey

Executive Marketing Manager: Deborah S. Yarnell

Marketing Manager: Katie Hall

Associate Marketing Manager: Mark Pierro

Print Buyer: Bev Breslin

Production Director: Wendy Troeger

Production Manager: Mark Bernard

Content Project Manager: Barbara LeFleur

Senior Art Director: Benj Gleeksman

For product information and technology assistance, contact us at
**Cengage Customer & Sales Support, 1-800-354-9706
or support.cengage.com.**

For permission to use material from this text or product, submit all requests online at **www.copyright.com**.

Library of Congress Control Number: 2007941007

ISBN-13: 978-1-4180-5389-5
ISBN-10: 1-4180-5389-9

Cengage
200 Pier 4 Boulevard
Boston, MA 02210
USA

Cengage is a leading provider of customized learning solutions with employees residing in nearly 40 different countries and sales in more than 125 countries around the world. Find your local representative at: **www.cengage.com**.

To learn more about Cengage platforms and services, register or access your online learning solution, or purchase materials for your course, visit **www.cengage.com**.

Notice to the Reader

Printed in the United States of America
Print Number: 06   Print Year: 2022

Contents

Preface

The *Student Workbook to Accompany Power Equipment Engine Technology* is designed to reinforce students' comprehension of the core textbook material, and to guide them through inspection, diagnostic, and service/repair procedures in the lab.

Each *Workbook* chapter is related to the content in the respective chapter in *Power Equipment Engine Technology*. The chapters include theory-based Shop Assignments and performance-based Job Sheets: The Shop Assignments are knowledge assessments that can be completed in the classroom, shop, or as homework assignments, while the Job Sheets offer step-by-step guidelines, checkpoints, and questions for hands-on maintenance. The *Workbook* chapters conclude with a Chapter Test—a mix of true/false, multiple choice, and short answer questions—designed to help measure students' content comprehension.

CHAPTER

1

Introduction to Power Equipment Engine Technology

Shop Assignment 1-1

Name: _____ Date: _____ Instructor: _____

Power Equipment Engine Industry Opportunities

Objective

After completing this activity sheet, you should be able to identify the employment opportunities available to you in the power equipment field and those that appeal to you the most.

Directions

From the list below, choose three jobs available in the power equipment field as identified in your textbook and list them in your personal order of preference.

General manager   *Salesperson*
Service writer   *Parts manager*
Technician   *Service manager*

1. List your top three choices:

    a. _____

    b. _____

    c. _____

2. Starting with your first choice, list the skills you will need to be successful in your chosen job title as described in your textbook.

    a. _____

    b. _____

    c. _____

    d. _____

3. Now list the skills you will need to be successful in the second job title you listed above.

a. _____

b. _____

c. _____

d. _____

4. Finally, list the skills needed for your third choice.

a. _____

b. _____

c. _____

d. _____

INSTRUCTOR VERIFICATION: _____

Shop Assignment 1-2

Name: _____ Date: _____ Instructor: _____

Creating a Five-Year Career Plan

Objective
After completing this activity sheet, you should have created a career plan that will guide you for the next five years.

Directions: Having a plan can mean the difference between success and failure. Create a plan that outlines where you want to be in your career in five years after graduation. Set goals for yourself and write down steps that will enable you to reach them. For instance, you might decide you want to own your own repair shop. Being a shop owner means more than just knowing how to repair generator engines. Consider what you will need to know and how you will acquire that knowledge.

1. One year:

   _____

   _____

   _____

   D0551653

2. Two years:

   _____

   _____

   _____

3. Three years:

   _____

   _____

   _____

4. Four years:

   _____

   _____

   _____

5. Five years:

   _____

   _____

   _____

INSTRUCTOR VERIFICATION: _____

Shop Assignment 1-2

Name: _____   Date _____   Instructor _____

Creating a Five-Year Career Plan

Objective

After completing this activity sheet, you should have created a career plan that will guide you for the next five years.

Directions: Having a plan can mean the difference between success and failure. Create a plan that outlines where you want to be in your career in five years after graduation. Set goals for yourself and write down steps that will enable you to reach them. For instance, you might decide you want to own your own repair shop. Being a shop owner means more than just knowing how to repair generator engines. Consider what you will need to know and how you will acquire that knowledge.

1. One year:

_____

_____

2. Two years:

_____

_____

_____

3. Three years:

_____

_____

_____

4. Four years:

_____

_____

_____

5. Five years:

_____

_____

_____

INSTRUCTOR VERIFICATION: _____

Shop Assignment 1-3

Name: _____ Date: _____ Instructor: _____

Equipment and Engine Training Council

Objective
After completing this activity sheet, you should be familiar with the Equipment and Engine Training Council (EETC) Web site and be able to identify 10 reasons why EETC certification is important to a career in power equipment repair.

Directions
Log on to the EETC site at http://www.eetc.org. Navigate to the "Technician Certification" page.

1. List the 10 reasons why you should be certified as specified on the EETC Web site.

   a. _____

   b. _____

   c. _____

   d. _____

   e. _____

   f. _____

   g. _____

   h. _____

   i. _____

   j. _____

2. Go to the "Industry Jobs" page. Click on the "Find a Job" tab and list the number of jobs listed on that page.

   a. Number of jobs found: _____

INSTRUCTOR VERIFICATION: _____

CHAPTER 1

Test

True/False

Indicate whether the statements are true or false.

_____ 1. External combustion steam engines are a direct descendant of internal combustion engines.

_____ 2. Historians credit James Watt with the invention of the first practical steam engine, in 1669.

_____ 3. The unit of electrical power called the watt was named after James Watt.

_____ 4. In the 19th century, gasoline was considered a waste product of oil.

_____ 5. Nikolas Otto made the gasoline engine a commercial success when he adopted the four-stroke cycle.

Multiple Choice

Identify the choice that best completes the statement or answers the question.

_____ 6. The goal of OPEESA is to _____.

    a. certify service technicians

    b. assist distributors in achieving outstanding channel performance

    c. create certification standards

    d. create curriculum for testing

_____ 7.   Early work on power equipment engines centered on _____.

    a.   building an engine to power farm equipment

    b.   building an engine to crush rocks

    c.   building an engine to pump water out of coal mines

    d.   building an engine to power an automobile

_____ 8.   The internal combustion engine was superior to the external combustion engine because it _____.

    a.   was lighter   c.   started quicker

    b.   was smaller   d.   all of the above

_____ 9.   _____ is responsible for writing up repair orders.

    a.   Lot attendant   c.   Engine technician

    b.   Service writer   d.   Setup technician

_____ 10.   The _____ holds the highest position in the service department.

    a.   general manager   c.   service manager

    b.   customer service   d.   service writer
       representative

INSTRUCTOR VERIFICATION: _____

CHAPTER 2

Safety First

Shop Assignment 2-1

Name: _____ Date: _____ Instructor: _____

Create a Shop Familiarization Floor Plan

Objective

After completing this activity sheet, you should be able to identify the location of important safety-related items in the shop.

Directions

Draw a simple floor plan of your shop and list the location of the items below:

Fire extinguisher	*Fuel dump*	*MSDS*	*Eyewash station*
Oil dump	*Station*	*Telephone*	*Shop towel disposal*

INSTRUCTOR VERIFICATION:

Safety First

Shop Assignment 2-1

Name: _____ Date: _____ Instructor: _____

Create a Shop Familiarization Floor Plan

Objective

After completing this activity sheet, you should be able to identify the location of important safety related items in the shop.

Directions

Draw a simple floor plan of your shop and list the location of the items below:

Fire extinguisher	Fuel dump	MSDS	Eyewash station
Oil dump	Station	Telephone	Shop towel disposal

INSTRUCTOR VERIFICATION: _____

Shop Assignment 2-2

Name: _____ Date: _____ Instructor: _____

Create a Fire Evacuation Plan

Objective

Having an evacuation plan can mean the difference between life and death when a fire occurs. Upon completion of this activity sheet, you should know the quickest route out of the facility in the event of a fire.

Directions

Create your plan now so you know what to do should you ever need to evacuate. Your plan should show an evacuation route and where you should meet once out of the building. Using your knowledge of the facility and its floor plan, draw a simple map showing where the nearest exit is located and the quickest route to get to it.

1. Draw the floor plan of your facility showing the route to the nearest exit.

2. Ask the instructor where your class should meet in case of an evacuation and list it below.

_____

INSTRUCTOR VERIFICATION: _____

Name _____ Date _____ Instructor _____

Create a Fire Evacuation Plan

Objective

Having an evacuation plan can mean the difference between life and death when a fire occurs. Upon completion of this activity sheet, you should know the quickest route out of the facility in the event of a fire.

Directions

Create a plan now so you know what to do should you ever need to evacuate. Your plan should show an evacuation route and where you should meet once out of the building. Using your knowledge of the facility and its floor plan, draw a simple map showing where the nearest exit is located and the quickest route to get to it.

1. Draw the floor plan of your facility showing the route to the nearest exit.

2. Ask the instructor where your class should meet in case of an evacuation and list it below.

_____

INSTRUCTOR VERIFICATION: _____

Shop Assignment 2-3

Name: _____ Date: _____ Instructor: _____

Create an Emergency Contact and Address Form

Objective
After completing this activity sheet, you should have a reference card that can be used if an emergency requires contacting next of kin or relative.

Directions
Create a form that gives the contact information for a relative in the event of an emergency.

1. Person who should be contacted in the event of an emergency: _____

2. Address: _____

3. Phone number: _____ Cell: _____

4. Relationship to you: _____

INSTRUCTOR VERIFICATION:

Shop Assignment 2-3

Name _____ Date: _____ Instructor: _____

Create an Emergency Contact and Address Form

Objective

After completing this activity sheet, you should have a reference card that can be used if an emergency requires contacting next of kin or relative.

Directions

Create a form that gives the contact information for a relative in the event of an emergency

1. Person who should be contacted in the event of an emergency: _____

2. Address: _____

3. Phone number: _____ Cell: _____

4. Relationship to you: _____

Shop Assignment 2-4

Name: _____ Date: _____ Instructor: _____

Locating Tools and Chemicals

Objective
After completing this activity sheet, you should be able to identify the location of special tools and chemicals commonly used in the shop.

Directions
Draw a simple floor plan of your shop and list the location of the items below.

Commonly used chemicals   *Fuel*   *Oil*   *Special tools*

INSTRUCTOR VERIFICATION:

Shop Assignment 2-5

Name: _____ Date: _____ Instructor: _____

Fire Safety

Objective

After completing this activity sheet, you should be able to identify the fire triangle, the four fire classes, and the proper use of a fire extinguisher.

Directions

Answer the questions below.

1. What three things make up the fire triangle?

   a. _____

   b. _____

   c. _____

2. There are four classes of fire: A, B, C, and D. Identify the class of fire created by each material by putting the letter of each type of material in the correct column. The first one is done for you.

	Class A	Class B	Class C	Class D
a. Paper	a	_____	_____	_____
b. Carburetor cleaner	_____	_____	_____	_____
c. Sales receipts	_____	_____	_____	_____
d. Dust covers	_____	_____	_____	_____
e. Cleaning cloths	_____	_____	_____	_____
f. Titanium shavings	_____	_____	_____	_____
g. Wiring insulation	_____	_____	_____	_____
h. Propane	_____	_____	_____	_____
i. Oily rags	_____	_____	_____	_____
j. Work aprons	_____	_____	_____	_____
k. Magnesium	_____	_____	_____	_____
l. Work area partitions	_____	_____	_____	_____
m. Gasoline	_____	_____	_____	_____
n. Electrical box	_____	_____	_____	_____

3. List the type of extinguisher you would use to fight each type of fire by writing the letter of the extinguisher type. You may use some choices more than once.

    a. CO_2 b. Foam   c. Halon   d. Dry powder   e. Water

    Class A _____   Class B _____   Class C _____   Class D _____

4. What is an incipient fire? *Circle the correct letter.*

    a. A fire that has flames less than two feet high.

    b. A fire that is smoldering but where no flames are showing.

    c. A fire started by arson.

5. What does PASS stand for?

    P _____

    A _____

    S _____

    S _____

6. You should aim the fire extinguisher at the **base/flames** of a fire. *Circle one.*

7. Take a look at the fire extinguishers in your shop and list the class of fire they will fight.

    _____   _____

8. What symbol is on the fire extinguisher in your shop? _____

INSTRUCTOR VERIFICATION:

Shop Assignment 2-6

Name: _____ Date: _____ Instructor: _____

Tool Safety

Objective
After completing this activity sheet, you should be able to identify the types of tools commonly used in the shop and their safety precautions.

Directions
Using your textbook as a resource, answer the questions below according to the guidelines set forth in Chapter 2. Please note that every scenario described actually happened.

1. Christine is trying to remove a bearing from a crankshaft but the bearing remover is broken, so she is using a screwdriver as a chisel. Why is this wrong and what is the solution?

   _____

   _____

   _____

2. Barney is using an impact driver and a regular socket to loosen a 13/16'' nut. Why is this wrong and what is the solution?

   _____

   _____

   _____

3. Tom's just bored a cylinder and needs to clean the metal chips off of the bore table but the vacuum is broken, so he is using an air blower to clean the bore table. Why is this wrong and what is the solution?

   _____

   _____

   _____

4. Roger is finishing up repairs on a generator and he needs a #2 point Phillips screwdriver. He doesn't have one close at hand, so he asks Bill to toss him one. Why is this wrong and what is the solution?

   _____

   _____

   _____

5. Steve dropped his hammer in the oil pan. The handle is coated with oil but he has to get the crankshaft bearings out before closing time, so he keeps working with it. Why is this wrong and what is the solution?

_____

_____

_____

6. Roger's cut-off wheel is just about worn out. The edges are frayed and it's past the minimum diameter recommended by the manufacturer but he only needs to make one more cut and he's done for the day, so he uses it. Why is this wrong and what is the solution?

_____

_____

_____

7. Anne finds that her drill motor won't turn unless she pushes the power cord into the drill with one hand while holding the drill with the other. Why is this wrong and what is the solution?

_____

_____

_____

8. Carlos needs to clean the chain on his customer's chain saw, so he fills a small glass with gasoline and uses a toothbrush to scrub it. Why is this wrong and what is the solution?

_____

_____

_____

9. Mike is using a high-speed air tool that makes a very high-pitched sound. He has been using this tool for several hours without ear protection. What's wrong here?

_____

_____

_____

10. Julian is drilling out a hole in a steel bar. He is using a drill press and holding the bar with his hand in case the drill bit catches. Why is this wrong and what is the solution?

_____

_____

_____

INSTRUCTOR VERIFICATION: _____

Shop Assignment 2-7

Name: _____ Date: _____ Instructor: _____

Using Personal Protective Equipment

Objective
After completing this activity sheet, you should be able to identify types of commonly used personal protective equipment (PPE) and their safety precautions.

Directions
Using your textbook as a resource, answer the questions below according to the guidelines set forth in Chapter 2. In some of the scenarios described below, you may find that there is more than one PPE needed to protect the technician. List all that are appropriate.

1. You have to paint a battery box damaged by battery acid. What PPE should you use?

   _____

2. Some service departments require that you watch out for yourself by wearing these at all times.

   _____

3. You are doing some arc welding. What PPE should you use?

   _____

4. You are using a high-speed cutting bit. What PPE should you use?

   _____

5. A battery explodes while on the charger and battery acid gets in your eyes. What PPE should you have used?

   _____

6. You need to remove a carburetor from the carburetor cleaning solution where it is resting at the bottom. What PPE should you use?

   _____

INSTRUCTOR VERIFICATION: _____

Shop Assignment 2-7

Name: _____ Date: _____ Instructor: _____

Using Personal Protective Equipment

Objective

After completing this activity sheet, you should be able to identify types of commonly used personal protective equipment (PPE) and their safety precautions.

Directions

Using your textbook as a resource, answer the questions below, according to the guidelines set forth in Chapter 2. In some of the scenarios described below, you may find that there is more than one PPE needed to protect the technician. List all that are appropriate.

1. You have to paint a battery box damaged by weather and acid. What PPE should you use?

_____

2. Some service departments require that you watch out for yourself by weather concerns at all times.

_____

3. You are doing some arc welding. What PPE should you use?

_____

4. You are using a high-speed cutting bit. What PPE should you use?

_____

5. A battery explodes while on the charger and battery acid gets in your eyes. What PPE should you have used?

_____

6. You need to remove a carburetor from the carburetor cleaning solution while it is resting at the bottom. What PPE should you use?

_____

INSTRUCTOR VERIFICATION: _____

CHAPTER 2

Test

True/False

Indicate whether the statements are true or false.

_____ 1. A Class A fire involves live electrical equipment.

_____ 2. A Class B fire involves flammable liquids.

_____ 3. The fire triangle includes oxygen.

_____ 4. Dry powder compounds and dry chemical extinguishers are the two primary methods to extinguish Class D fires.

_____ 5. Batteries can emit highly explosive gasses.

Multiple Choice

Identify the choice that best completes the statement or answers the question.

_____ 6. _____ is a poisonous gas that is a by-product of combustion.

    a. Carbon monoxide   c. Argon

    b. Nitrogen   d. Methane

_____ 7. _____ prevent injury to your instep and feet.

    a. Metatarsal guards   c. Open-toed shoes

    b. Steel-toed boots   d. Sneakers

_____   8.   _____ is a federal agency that enforces safety standards in repair shops and other businesses.

    a. NFPA   c. OSHA

    b. NEC   d. MIC

_____   9.   _____ is a potential source of ignition that you should recognize.

    a. AC   c. Carbon dioxide

    b. Spontaneous combustion   d. Carbon monoxide

_____   10.   _____ is a highly effective and oxygen-depleting gas used to extinguish all types of fires.

    a. Zenon   c. Halon

    b. Oxygen   d. Argon

Short Answers

11.   What does the acronym PASS stand for and why is it important?

12.   What does the acronym MSDS stand for and why is it important?

13.   Debilitating back injuries are common among technicians. List three ways you can prevent becoming a victim of back injury.

14.   Noise-induced hearing loss is permanent. How can you prevent hearing loss in the workplace?

15.   What personal protection equipment should be used in a power equipment service department?

INSTRUCTOR VERIFICATION:

Tools

Shop Assignment 3-1

Name: _____ Date: _____ Instructor: _____

Tool Knowledge Assessment

Objective
By completing this activity sheet, you should be able to demonstrate basic tool knowledge.

Directions
After reading the textbook chapter on hand tools, answer the questions below to the best of your ability.

A. How does a Reed & Prince tip differ in appearance from a Phillips tip?

_____

B. Why isn't the adjustable wrench used more often in engine repair?

_____

_____

C. What is a flare nut wrench used for?

_____

D. Why use a 6-point wrench when a 12-point is available?

_____

E. When would you use a dead blow hammer?

_____

F. What is the difference between a tap and a die?

_____

G. What is a torque wrench used for?

_____

H. What is a breaker bar for?

_____

I. When is a 12-point socket superior to a 6-point?

_____

J. What are the three common drive sizes for sockets?

_____

INSTRUCTOR VERIFICATION: _____

Shop Assignment 3-2

Name: _____ Date: _____ Instructor: _____

Precision Measuring Tools

Objective
After reading Chapter 3, you should be able to answer the questions below relating to the use of the precision measuring tools and identify them from photographs and drawings.

Directions
Using your textbook as a resource, answer the questions below.

1. Name nine parts of a micrometer.

   _____

   _____

2. Vernier calipers can make three types of measurements. What are they?

   _____

3. Which is more accurate, a micrometer or a digital caliper?

   _____

4. What is the dial indicator used to measure?

   _____

5. Name two types of torque wrenches commonly found in a repair shop.

   _____

6. What is a feeler gauge used to measure?

   _____

INSTRUCTOR VERIFICATION: _____

Shop Assignment 3-2

Name: _____ Date: _____ Instructor: _____

Precision Measuring Tools

Objective

After reading Chapter 3, you should be able to answer the questions below relating to the use of the precision measuring tools and identify them from photographs and drawings.

Directions

Using your textbook as a resource, answer the questions below.

1. Name nine parts of a micrometer.

   _____

2. Vernier calipers can make three types of measurements. What are they?

   _____

3. Which is more accurate, a micrometer or a digital caliper?

   _____

4. What is the dial indicator used to measure?

   _____

5. Name two types of torque wrenches commonly found in a repair shop.

   _____

6. What is a feeler gauge used to measure?

   _____

INSTRUCTOR VERIFICATION: _____

Shop Assignment 3-3

Name: _____ Date: _____ Instructor: _____

Test Instruments

Objective
After reading Chapter 3, you should be able to answer the questions below.

Directions
Using your textbook as a resource, fill in the blanks.

1. The most common electrical testing tool is _____.

2. A timing light is used to check an engine's _____ system.

3. You would use a _____ to check cylinder pressure.

INSTRUCTOR VERIFICATION:

Shop Assignment 3-3

Name _____ Date _____ Instructor: _____

Test Instruments

Objective

After reading Chapter 3, you should be able to answer the questions below.

Directions

Using your textbook as a resource, fill in the blanks.

1. The most common electrical testing tool is _____.

2. A timing light is used to check an engine's _____ system.

3. You would use a _____ to check cylinder pressure.

INSTRUCTOR VERIFICATION: _____

CHAPTER 3 Test

True/False

Indicate whether the statement is true or false.

_____ 1. The size of a wrench is determined by the width of the opening at the end of the wrench.

_____ 2. Use of an extension on an open-end wrench would provide more torque.

_____ 3. A flare nut wrench contacts the flare nut at 360°.

_____ 4. A 6-point wrench is thicker than a 12-point wrench.

_____ 5. Adjustable wrenches can slip and round off corners of bolt heads.

Multiple Choice

Identify the choice that best completes the statement or answers the question.

_____ 6. A(n) _____ is made from a stronger grade of steel.
   a. Impact socket   c. Torx
   b. 12-point wrench   d. Allen

_____ 7. A(n) _____ is star shaped.
   a. Reed & Prince screwdriver   c. Torx
   b. Allen wrench   d. Pozidrive screwdriver

_____   8.   A(n) _____ is filled with lead shot to prevent bouncing.

      a.   ball peen hammer   c.   mallet

      b.   impact driver   d.   dead blow hammer

_____   9.   A _____ cuts female threads.

      a.   die   c.   thread file

      b.   tap   d.   nut splitter

_____ 10.   A(n) _____ removes broken bolts.

      a.   tap   c.   drill press

      b.   impact wrench   d.   screw extractor

Short Answers

11.   What are the guidelines expressed in the textbook for using power drills?

12.   Why should you exercise caution when using a bench grinder?

13.   What is a puller used for?

14.   What are the three types of measurements a vernier caliper can make?

15.   Define anvil, spindle, sleeve, and thimble as these terms apply to a micrometer.

INSTRUCTOR VERIFICATION:

CHAPTER 4

Measuring Systems, Fasteners, and Thread Repair

Shop Assignment 4-1

Name: _____ Date: _____ Instructor: _____

Fasteners, Measuring Systems, and Thread Repair Knowledge Assessment

Objective

After completing this activity sheet, you should be able to identify the two different measuring systems used in power equipment engines and identify the fastener types used on these machines.

Directions

Using your textbook as a resource, answer the questions below.

1. Name the two types of measuring systems used in power equipment engines.

   a. _____   b. _____

2. What is the base point measurement of the metric system? _____

3. How many millimeters are there in a meter? _____

4. What is the technical term used to describe the tension caused by tightening the fastener that holds two parts together? _____

5. List the four critical bolt dimensions.   1. _____   2. _____   3. _____   4. _____

6. If someone asked you to get a 10-mm bolt, would you get him or her a bolt that had a 10-mm **head** or a 10-mm **diameter**? *Circle the answer.*

7. You notice that a bolt head is stamped with the letter "L." What does this mean?

    _____

8. A bolt head is marked with the number "8.8." Is this bolt **stronger** or **weaker** than one with "7" stamped on the head? *Circle the answer.*

9. Why are castle nuts shaped like they are? _____

10. Technician A says that he often substitutes a stronger bolt than originally used by the manufacturer. Technician B says stronger bolts are brittle and may break if not used in the correct application. Who is right?

    A   B   *Circle the answer*

11. What tool should you use to avoid overstretching a bolt when tightening it?

    _____

12. When disassembling a component with various fastener sizes, which ones do you loosen first?

    _____

13. Lubricating fasteners is always a good idea.

    True   False   *Circle the answer*

INSTRUCTOR VERIFICATION:

Shop Assignment 4-2

Name: _____ Date: _____ Instructor: _____

Fastener Anatomy and Identification

Objective

After completing this activity sheet, you should be able to demonstrate your knowledge of the anatomy of fastener types and identify specific types of fasteners used in the power equipment engine field.

Directions

Using your textbook as a resource, answer the questions below.

1. Identify the basic parts of a fastener by marking up the drawing at the right. Indicate the location and dimensions of the following:

   Head

   Length

   Thread pitch

   Nominal diameter

   A.
   U.S. Customary

2. According to your textbook, where would you use a lock plate?

   a. _____

   b. _____

   c. _____

INSTRUCTOR VERIFICATION: _____

Shop Assignment 4-2

Name _____   Date: _____   Instructor: _____

Fastener Anatomy and Identification

Objective

After completing this activity sheet, you should be able to demonstrate your knowledge of the anatomy of fastener types and identify specific types of fasteners used in the power equipment engine field.

Directions

Using your textbook as a resource, answer the questions below.

1. Identify the basic parts of a fastener by matching up the drawing at the right. Indicate the location and dimensions of the following:

A
U.S. Customary

Head _____

Length _____

Thread pitch _____

Nominal diameter _____

2. According to your textbook, where would you use a lock plate?

a. _____

b. _____

c. _____

Job Sheet 4-1

Name: _____ Date: _____ Instructor: _____

Using a Torque Wrench

Objective
After completing this job sheet, you should be able to correctly use a torque wrench.

Directions
Engines often have assemblies held together with different-sized fasteners. Use the service manual to identify the amount of torque needed for each fastener and the order in which the fasteners are torqued.

Tools and Equipment
Torque wrench, safety glasses, cylinder head or equivalent, service manual

1. Loosen all bolts in preparation for torquing. Check each one to make sure the threads are not damaged before starting.

2. Does the service manual specify that the threads should be oiled?

3. List the sizes of fasteners and the amount of torque each size requires:

   a. _____   _____

   b. _____   _____

   c. _____   _____

INSTRUCTOR VERIFICATION:

Job Sheet 4-1

Name: _____ Date: _____ Instructor: _____

Using a Torque Wrench

Objective

After completing this job sheet, you should be able to correctly use a torque wrench.

Directions

Engines often have assemblies held together with different-sized fasteners. Use the service manual to identify the amount of torque needed for each fastener and the chart in which the fasteners are torqued.

Tools and Equipment

Torque wrench, safety glasses, cylinder head or equivalent, service manual

1. Loosen all bolts in preparation for torquing. Check each one to make sure the threads are not damaged before starting.

2. Does the service manual specify that the threads should be oiled? _____

3. List the sizes of fasteners and the amount of torque each size requires:

a. _____

b. _____

c. _____

INSTRUCTOR VERIFICATION: _____

Job Sheet 4-2

Name: _____ Date: _____ Instructor: _____

Thread Repair

Objective

After completing this job sheet, you should be able to remove broken fasteners and repair damaged threads.

Directions

Your instructor will provide you with the necessary training aids to complete this job sheet. You will purposely over-torque a fastener until it breaks, then you will use a screw extractor to remove it.

Tools and Equipment

6-mm bolt, aluminum block with threaded holes, screw extractor, drill motor, drill bit

1. Take a 6-mm bolt and thread it into one of the holes of the thread repair training aid. Tighten it with a wrench until it breaks off, then follow the procedure explained in the textbook for using a screw extractor to remove it.

2. Show the broken bolt to the instructor before you remove it.

3. Once the bolt is removed, have your instructor check your work.

INSTRUCTOR VERIFICATION: _____

Job Sheet 4-3

Name: _____ Date: _____ Instructor: _____

Thread Insert

Objective

After completing this job sheet, you should be able to properly install a thread insert. The thread insert will be installed so that it is flush with the top surface of the aluminum block and the insert will easily accept a new bolt. You will be able to torque the bolt to 8 ft-lb.

Directions

Your instructor will provide you with the necessary training aids to complete this job sheet. You will install a thread insert following the instructions provided with the thread insert kit.

Tools and Equipment

Aluminum block with threaded holes, 6-mm thread insert kit, drill motor or drill press, tap handle

1. Follow the instructions provided with the thread repair kit and install it in the provided aluminum block.

2. After the thread insert is installed, thread a new bolt into the insert and torque it to 8 ft-lb.

INSTRUCTOR VERIFICATION: _____

Job Sheet 4-3

Name: _____ Date: _____ Instructor: _____

Thread Insert

Objective

After completing this job sheet, you should be able to properly install a thread insert. The thread insert will be installed so that it is flush with the top surface of the aluminum block and the insert will easily accept a new bolt. You will be able to torque the bolt to 8 ft-lb.

Directions

Your instructor will provide you with the necessary training aids to complete this job sheet. You will install a thread insert following the instructions provided with the thread insert kit.

Tools and Equipment

Aluminum block with threaded holes, 6-mm thread insert kit, drill motor or drill press, tap handle

1. Follow the instructions provided with the thread repair kit and install it in the provided aluminum block.

2. After the thread insert is installed, thread a new bolt into the insert and torque it to 8 ft-lb.

INSTRUCTOR VERIFICATION. _____

CHAPTER 4

Test

True/False

Indicate whether the statement is true or false.

_____ 1. A millimeter equals 1/10,000 of a meter.

_____ 2. A centimeter equals 10 millimeters.

_____ 3. Preload is the technical term for the tension caused by tightening a fastener.

_____ 4. ISO stands for Industry Standards Organization.

_____ 5. Bolts come in left- and right-hand threads.

Multiple Choice

Identify the choice that best completes the statement or answers the question.

_____ 6. A _____ bolt is designed to resist loosening.

     a. DR   c. Torx

     b. UBS   d. CT

_____ 7. A _____ bolt is self-tapping.

     a. DR   c. Torx

     b. UBS   d. CT

_____ 8. A _____ bolt head is patented to carry a greater amount of torque from the socket to the bolt.

　　a. DR　　　　　　　　　　　c. Torx

　　b. UBS　　　　　　　　　　d. CT

_____ 9. A(n) _____ nut allows a cotter pin to be used.

　　a. castle-headed　　　　　c. acorn

　　b. well　　　　　　　　　　d. stake

_____ 10. A(n) _____ nut is rubber with a brass sleeve in the center.

　　a. castle-headed　　　　　c. acorn

　　b. well　　　　　　　　　　d. stake

Short Answers

11. What is the difference between a cone washer and a split-ring washer?

12. Describe the procedure for preparing a fastener for a thread-locking compound.

13. Define axial tension.

14. Define torsion.

15. Explain "plastic region" as applied to a bolt.

INSTRUCTOR VERIFICATION:

Basic Engine Operation and Configurations

Shop Assignment 5-1

Name: _____ Date: _____ Instructor: _____

Basic Engine Operation and Configuration Knowledge Assessment

Objective

By correctly completing this assessment, you will demonstrate an understanding of basic two- and four-stroke engine operation.

Directions

Answer the questions below using your textbook as a resource.

1. This engine gives a power stroke every 720°. Is it a **two**- or **four**-stroke engine? *Circle the answer.*

2. This engine has no valves. Is it a **two**- or **four**-stroke engine? *Circle the answer*.

3. This engine is environmentally friendly. Is it a **two**- or **four**-stroke engine? *Circle the answer.*

4. The power stroke is an upstroke. **True** or **false**? *Circle the answer.*

5. This device measures an engine's torque and horsepower. _____

6. This engine uses the piston to time when the exhaust exits the combustion chamber. Is it a **two**-or **four**-stroke engine? *Circle the answer.*

7. An engine's displacement is determined by the length of its stroke and the diameter of its piston. **True** or **false**? *Circle the answer.*

8. Torque is a measure of an engine's ability to turn its crankshaft. **True** or **false**? *Circle the answer.*

9. Compression ratio is a comparison of combustion chamber volumes at BDC and TDC. **True** or **false**? *Circle the answer.*

10. An engine with an 8.91:1 compression ratio will make more power than one with a 10.0:1 ratio. **True** or **false**? *Circle the answer.*

11. This is a crankshaft from a two-stoke engine. **True** or **false**? *Circle the answer.*

12. This is the connecting rod and piston from a four-stroke engine. **True** or **false**? *Circle the answer.*

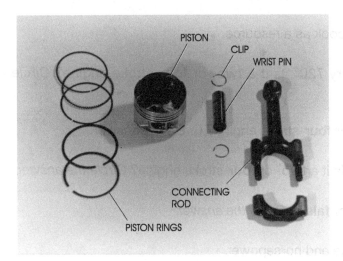

13. A 500-cc engine has roughly the same displacement as a 45 cu. in engine. **True** or **false**? *Circle the answer*.

14. Who invented the calculation for engine horsepower? _____

INSTRUCTOR VERIFICATION:

Shop Assignment 5-2

Name: _____ Date: _____ Instructor: _____

Identifying Engine Configurations

Objective

After correctly completing this assessment, you should be able to identify the two common types of engines used in power equipment applications.

Directions

Answer the questions below using your textbook as a resource. Write the type of engine configuration under each photo.

_____   _____

INSTRUCTOR VERIFICATION:

Shop Assignment 5-3

Name: _____ Date: _____ Instructor: _____

Calculating Displacement

Objective

After correctly completing this assessment, you should be able to calculate engine displacement in both metric and SAE systems.

Directions

Answer the questions below using your textbook as a resource. Use the following formula: Displacement = $B \times B \times 0.7854 \times S \times N$.

1. A twin cylinder engine has a bore of 52 mm and stroke of 48 mm. Calculate its displacement.
   _____cc

2. Use the same bore and stroke numbers as above, but this time make it a four cylinder engine.
   _____cc

3. A single cylinder engine has a stroke of 3.5" and a bore of 3.0". What is its displacement?
   _____"

INSTRUCTOR VERIFICATION: _____

Name _____   Date: _____   Instructor: _____

Calculating Displacement

Objective

After correctly completing this assessment, you should be able to calculate engine displacement in both metric and SAE systems.

Directions

Answer the questions below using your textbook as a resource. Use the following formula:
Displacement = B × B × 0.7854 × S × N.

1. A twin cylinder engine has a bore of 57 mm and stroke of 48 mm. Calculate its displacement.

_____ cc

2. Use the same bore and stroke numbers as above, but this time make it a four cylinder engine.

_____ cc

3. A single cylinder engine has a stroke of 3.5" and a bore of 3.0". What is its displacement?

_____ "

CHAPTER 5

Test

True/False

Indicate whether the statement is true or false.

_____ 1. The basic operation of a four-stroke engine is divided into four events.

_____ 2. A four-stroke engine has a power stroke every 180°.

_____ 3. A crankshaft turns reciprocating motion into rotary motion.

_____ 4. The piston rises on the exhaust stroke of a four-stroke engine.

_____ 5. The fuel mixture enters the combustion chamber from the area around the crankshaft of a two-stroke engine.

Multiple Choice

Identify the choice that best completes the statement or answers the question.

_____ 6. Torque is measured in _____.

    a. ft-lb   c. ft-lb/min

    b. ft-lb/s   d. kilopascals

_____ 7. Engine displacement is the volume of space that the _____ moves.

    a. crankshaft   c. camshaft

    b. piston   d. connecting rod

_____ 8. Displacement can be measured in cubic centimeters (cc) or _____.

        a. newton meters   c. millimeters squared

        b. degrees Kelvin   d. cubic inches

_____ 9. _____ is the comparison of the volume of the cylinder to the combustion chamber.

        a. Cylinder displacement   c. Engine efficiency

        b. Compression ratio   d. Brake horsepower

_____ 10. The _____ is supported on either end by bearings.

        a. crankshaft   c. clutch

        b. wrist pin clip   d. combustion chamber

Short Answers

11. What is a counter-balancer?

12. How does the fuel–air mixture get from the bottom of the crankcase to the combustion chamber in a two-stroke engine?

13. What happens in a two-stroke engine at the same time the piston is moving up on the compression stroke?

14. Explain how horsepower and torque are measured.

15. What is forced draft cooling?

INSTRUCTOR VERIFICATION:

Internal Combustion Engines

Shop Assignment 6-1

Name: _____ Date: _____ Instructor: _____

Internal Combustion Engine Knowledge Assessment

Objective
By correctly answering the questions below, you will demonstrate your knowledge of internal combustion engines.

Directions
Using your textbook as a resource, answer the questions below.

1. According to your textbook, there are seven scientific terms and principles associated with the operation of an internal combustion engine. Name them.

   1. _____
   2. _____
   3. _____
   4. _____
   5. _____
   6. _____
   7. _____

2. Using the terms and principles you named in Question 1, use a single word to describe how each of the seven principles or terms relates to an internal combustion engine. There may be more than one correct answer for some of them.

   Flywheel: _____

   Carburetor: _____

Engine oil: _____

Compression: _____

Combustion: _____

Battery: _____

3. Name the four stages of engine operation in order of occurrence.

    a. _____

    b. _____

    c. _____

    d. _____

4. What does a poppet valve do in a four-stroke engine?

    _____

5. What accomplishes the function of a poppet valve in a two-stroke engine?

    _____

6. What is stellite, and where is it used?

    _____

7. Refer to the drawing and name the parts of a valve.

    a. _____

    b. _____

    c. _____

    d. _____

    e. _____

    f. _____

    g. _____

8. Where does the valve actually contact the combustion chamber?

    _____

9. Name two common types of valve-opening devices used in power equipment engines.

    a. _____

    b. _____

    c. _____

10. What do the terms *ramp*, *duration*, and *base circle* refer to?

_____

_____

_____

11. Name two ways a camshaft can be actuated.

_____

_____

12. What terms describe when both the intake and exhaust valves are open at the same time?

_____

13. At what stage of engine operation does it occur?

_____

14. How many rings are there on a typical four-stroke piston?

_____

15. How many rings are there on a typical two-stroke piston?

_____

16. Where would you find a 120° crankshaft?

_____

17. At what stage does fuel induction occur?

_____

18. When does a spark plug fire?

_____

19. What does the transfer port do in a two-stroke engine?

_____

20. What does a labyrinth seal do on a two-stroke engine?

_____

INSTRUCTOR VERIFICATION: _____

Job Sheet 6-1

Name: _____ Date: _____ Instructor: _____

Top End Engine Inspection

Objective

After completing this job sheet, you should be able to disassemble the top end of a two- or four-stroke engine, identify its major components, and then reassemble it correctly.

Directions

Using a training aid designated for this project by your instructor, consult the appropriate service manual for instructions on disassembling the engine's top end, answer the questions, and then reassemble the engine.

1. Does this engine have poppet valves?

   Yes **No** *Circle the answer*

2. If you answered yes, how are the valves opened?

   Rocker arm **lifter** **does not apply** *Circle the answer*

3. How many rings are there on the piston?

   1 2 3 *Circle the answer*

4. Does the wrist pin ride in a bearing?

   Yes **No** *Circle the answer*

5. Does the piston have holes in the skirt?

   Yes **No** *Circle the answer*

6. Does the combustion chamber have a squish band?

   Yes **No** *Circle the answer*

7. Is the cylinder **cast iron** or **plated**? *Circle the answer*

8. Is the cylinder crosshatched?

   Yes **No** *Circle the answer*

9. What kind of top piston ring does the piston have?

   Standard **Keystone** **Dykes chrome plated** *Circle the answer*

10. Without removing the crankshaft, determine if it is a **single-** or **multi-piece** unit.
    Circle the answer.

Reassemble the engine according to the instructions in the service manual. Once the engine is reassembled, carefully turn the crankshaft two complete revolutions. If the engine will not turn past a certain point, stop immediately before you damage it. Call your instructor for further instructions.

INSTRUCTOR VERIFICATION: _____

CHAPTER

6

Test

True/False

Indicate whether the statement is true or false.

_____ 1. Air density increases as altitude increases.

_____ 2. The air we breathe is mostly oxygen.

_____ 3. Two-stroke engines use one-piece rods.

_____ 4. The intake event has the longest port duration of all two-stroke events.

_____ 5. Valve overlap is the time when both valves are closed.

Multiple Choice

Identify the choice that best completes the statement or answers the question.

_____ 6. During the _____ stage, the piston rises and compresses the air–fuel mixture.

    a. intake   c. power

    b. compression   d. exhaust

_____ 7. During the _____ stage, the air–fuel mixture is ignited.

    a. intake   c. power

    b. compression   d. exhaust

_____ 8. A _____ controls the gases coming into and out of a four-stroke engine's combustion chamber.

    a. reed valve   c. transfer port

    b. poppet valve   d. disc valve

_____ 9. _____ result(s) from unburned or raw fuel.

    a. Carbon dioxide   c. Hydrocarbons

    b. Oxides of nitrogen   d. Carbon monoxide

_____ 10. _____result(s) from partially burned fuel.

    a. Carbon dioxide   c. Hydrocarbons

    b. Oxides of nitrogen   d. Carbon monoxide

Short Answers

11. What are valve seats?

12. What is a rocker arm and what is it used for?

13. Define the term *duration*.

14. Why are pistons cam ground?

15. What is counterweight?

INSTRUCTOR VERIFICATION:

Lubrication and Cooling Systems

Shop Assignment 7-1

Name: _____ Date: _____ Instructor: _____

Lubrication and Cooling System Operation Knowledge Assessment

Objective

By correctly completing this assessment, you will demonstrate your knowledge of cooling and lubrication system concepts and principles.

Directions

Answer the questions below using your textbook as a resource.

1. Oil performs four functions in a power equipment engine. Name them.

    a. _____

    b. _____

    c. _____

    d. _____

2. API classifies oil by type. What is the highest API rating? _____

3. What does the "W" rating mean in the API ratings?

    _____

4. What is the viscosity index?

    _____

5. Name the two types of loads a ball bearing can carry.

   a. _____

   b. _____

6. Where would you find plain or precision insert bearings located in a power equipment engine?

   _____

7. How does a two-stroke engine get its lubrication?

   _____

8. What is the typical two-stoke premix ratio?

   _____

9. What does wet sump mean?

   _____

   _____

10. Name three types of oil pumps.

    a. _____

    b. _____

    c. _____

11. When does the oil pressure relief valve operate?

    _____

12. Name the three types of cooling systems.

    a. _____

    b. _____

    c. _____

13. What does the presence of coolant at the telltale hole indicate?

    _____

14. What is the checking pressure for a radiator cap? _____

15. What is the recommended coolant-to-water ratio? _____

16. How can you check the integrity of a cooling system?

    _____

17. Which type of oil pump is most common in modern engines? _____

18. How do you test a thermostat?

_____

19. Why would you use a hydrometer to check a cooling system?

_____

20. Power equipment engine manufacturers provide several ways for checking the oil level. Name two of them.

_____

INSTRUCTOR VERIFICATION: _____

Shop Assignment 7-2

Name: _____ Date: _____ Instructor: _____

Calculating Premix Ratios

Objective

By correctly completing this assessment, you should be able to demonstrate your ability to mix fuel and oil at the correct ratio.

Directions

Answer the questions below.

1. You have a lot of work to do this weekend and you are taking 10 gallons of gas. How much oil will you need to add to your gas for a 20:1 premix ratio?

   _____

2. You have 1 gallon of gasoline and your engine runs best on a premix ratio of 32:1. How much oil will you add to the gas?

   _____

3. You have 3 gallons of gas left. How much oil will you need to add to this gas for a 40:1 ratio?

   _____

4. You have 7.28 ounces of oil left and want to empty the bottle. How much gas will you need to add to the oil for a 50:1 ratio?

   _____

INSTRUCTOR VERIFICATION:

Name: _____ Date: _____ Instructor: _____

Calculating Premix Ratios

Objective
By correctly completing this assessment, you should be able to demonstrate your ability to mix fuel and oil at the correct ratio.

Directions
Answer the questions below.

1. You have a lot of work to do this weekend and you are taking 10 gallons of gas. How much oil will you need to add to your gas for a 25:1 premix ratio?

_____

2. You have 1 gallon of gasoline and your engine runs best on a premix ratio of 22:1. How much oil will you add to it a gas?

_____

3. You have 3 gallons of gas left. How much oil will you need to add to this gas for a 40:1 ratio?

_____

4. You have 7.25 ounces of oil left and want to empty the bottle. How much gas will you need to add to the oil for a 50:1 ratio?

_____

INSTRUCTOR VERIFICATION

Shop Assignment 7-3

Name: _____ Date: _____ Instructor: _____

Bearing Identification

Objective

After completing this shop assignment, you should be able to identify the different types of bearings and where they are most commonly used in a power equipment engine.

Directions

Write what types of bearings are pictured and where they might be used in a power equipment engine.

A

B

Type of bearing:

_____

Applications:

_____

_____

_____

Type of bearing:

_____

Applications:

_____

_____

_____

C

Type of bearing:

_____

Applications:

_____

_____

D

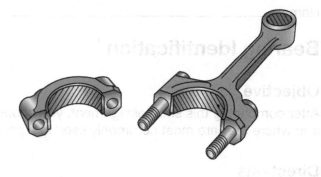

Type of bearing:

_____

Applications:

_____

_____

INSTRUCTOR VERIFICATION: _____

Job Sheet 7-1

Name: _____ Date: _____ Instructor: _____

Oil Pressure Check

Objective

After completing this job sheet, you should be able to correctly test the oil pressure of an engine with plain bearings.

Directions

Use the instructions provided in the appropriate service manual to check oil pressure.

Tools needed: Oil pressure gauge, service manual

1. Using a power equipment engine designated for this task by the instructor, remove the inspection plug and attach the oil pressure gauge.

2. Start the engine and observe the pressure reading.

    a. Observed oil pressure: _____

    b. Service manual specification: _____

    c. Did the engine meet specification?

    Yes   **No**

3. Rev the engine up to 3,000 rpm and note any increase in pressure.

    a. Oil pressure at 3,000 rpm: _____

4. What prevents the oil pressure from going too high when the engine revs up?

    _____

5. Remove the oil pressure gauge and reinstall the inspection plug. Make the vehicle ready for inspection by the instructor.

    _____

INSTRUCTOR VERIFICATION: _____

Lab Sheet 7-1

Name: _____ Date: _____ Instructor: _____

Oil Pressure Check

Objective

After completing this job sheet, you should be able to correctly test the oil pressure of an engine with plain bearings.

Directions

Use the instructions provided in the appropriate service manual to check oil pressure.

Tools needed: Oil pressure gauge, service manual

1. Using a power equipment engine designated for this task by the instructor, remove the inspection plug and attach the oil pressure gauge.

2. Start the engine and observe the pressure reading.

   a. Observed oil pressure: _____

   b. Service manual specification: _____

   c. Did the engine meet specifications?

      Yes _____ No _____

3. Rev the engine up to 3,000 rpm and note any increase in pressure.

   a. Oil pressure at 3,000 rpm: _____

4. What prevents the oil pressure from going too high when the engine revs up?

_____

5. Remove the oil pressure gauge and reinstall the inspection plug. Make the engine ready for inspection by the instructor.

_____

INSTRUCTOR VERIFICATION: _____

Job Sheet 7-2

Name: _____ Date: _____ Instructor: _____

Radiator Cap Inspection

Objective

After completing this job sheet, you should be able to correctly test the radiator cap.

Directions

Use the instructions provided in the appropriate service manual to check the radiator cap.

Tools needed: Cooling system pressure pump, cap adapter, service manual

1. A radiator cap must vent at the proper pressure or risk damaging the cooling system. Obtain a
    pressure test kit from the instructor and note the pressure listed on the cap.

    Cap pressure: _____

2. Attach the pressure pump to the cap and note the pressure at which the cap vents.

    Observed cap pressure: _____

3. Did this cap pass the test?

    Yes   **No**   *Circle the answer*

4. If a cap vents at too low a pressure, how will this affect the cooling system?

    _____

    _____

5. If you used a running vehicle for this test, replace the cap and make sure the cooling system is
    topped off. Prepare the engine for final inspection by the instructor.

    _____

INSTRUCTOR VERIFICATION: _____

Job Sheet 7-3

Name: _____ Date: _____ Instructor: _____

Cooling System Pressure Check

Objective
After completing this job sheet, you should be able to correctly test the integrity of a cooling system.

Directions
Use the instructions provided in the appropriate service manual to check the cooling system.

Tools needed: Cooling system pressure pump, radiator adapter, service manual

1. Remove the radiator cap from a cold vehicle and attach the pressure pump to the radiator.

2. What is the recommended test pressure for this engine? _____

3. Pump up the system to the recommended pressure and wait 1 minute. Check all hoses for leakage.

   a. Did the system hold pressure for 1 minute?

   Yes   **No**   *Circle the answer*

   b. Did you note any leakage?

   Yes   **No**   *Circle the answer*

   c. A system that does not hold pressure but shows no signs of external leakage may have a blown _____.

4. Replace the radiator cap and prepare the engine for final inspection.

INSTRUCTOR VERIFICATION: _____

Job Sheet 7-3

Name: _____ Date: _____ Instructor: _____

Cooling System Pressure Check

Objective

After completing this job sheet, you should be able to correctly test the integrity of a cooling system.

Directions

Use the instructions provided in the appropriate service manual to check the cooling system.

Tools needed: Cooling system pressure pump, radiator adapter, service manual

1. Remove the radiator cap from a cold vehicle and attach the pressure pump to the radiator.

2. What is the recommended test pressure for this engine? _____

3. Pump up the system to the recommended pressure and wait 1 minute. Check all hoses for leakage.

   a. Did the system hold pressure for 1 minute?

      Yes   No   Circle the answer

   b. Did you note any leakage?

      Yes   No   Circle the answer

   c. A system that does not hold pressure but shows no signs of external leakage may have a _____ blown.

4. Replace the radiator cap and prepare the engine for final inspection.

INSTRUCTOR VERIFICATION: _____

Job Sheet 7-4

Name: _____ Date: _____ Instructor: _____

Coolant Ratio Inspection

Objective

After completing this job sheet, you should be able to correctly test the ratio of coolant.

Directions

Obtain a hydrometer designated for this task by the instructor. Remove a sample of the coolant and note its strength.

Tools needed: Hydrometer

1. Remove the radiator cap from a cold engine and use the hydrometer to suck up a sample of the coolant.

2. Some hydrometers read in specific gravity and others read in degrees of protection from freezing or boil over. Which one does your tester read?

    _____

3. Does this coolant have sufficient ethylene glycol to protect it from freezing or boil over?

    Yes   **No**   *Circle the answer*

4. Replace the coolant you removed and reinstall the radiator cap. Prepare the engine for final inspection by the instructor.

INSTRUCTOR VERIFICATION: _____

Job Sheet 7-4

Name: _____ Date: _____ Instructor: _____

Coolant Ratio Inspection

Objective

After completing this job sheet, you should be able to correctly test the ratio of coolant.

Directions

Obtain a hydrometer designated for this task by the instructor. Remove a sample of the coolant and note its strength.

Tools needed: Hydrometer

1. Remove the radiator cap from a cold engine and use the hydrometer to suck up a sample of the coolant.

2. Some hydrometers read in specific gravity and others read in degrees of protection from freezing or boil over. Which one does your tester read?

   _____

3. Does this coolant have sufficient ethylene glycol to protect it from freezing or boil over?

   Yes _____ No _____ Circle the answer

4. Replace the coolant you removed and reinstall the radiator cap. Prepare the engine for final inspection by the instructor.

CHAPTER

7

Test

True/False

Indicate whether the statement is true or false.

_____ 1. Friction is the resistance to motion created when two surfaces move against each other.

_____ 2. As engine oil is moving around and through the engine, it is also cleaning the engine's parts.

_____ 3. Synthetic oils can operate over a wider temperature range than petroleum-based oils.

_____ 4. The highest classification of oil is SL.

_____ 5. The "W" in 10W40 stands for winter.

Multiple Choice

Identify the choice that best completes the statement or answers the question.

_____ 6. _____ bearings are the most popular because they provide the greatest amount of friction reduction and can handle axial and radial loads.

    a. Tapered   c. Needle

    b. Ball   d. Roller

_____ 7. _____ bearings use cylindrical shaped rollers.

    a. Tapered   c. Needle

    b. Ball   d. Roller

_____ 8. _____ ounces of oil are required to produce 5 gallons of premix at a 40:1 ratio.

   a. 8   c. 14

   b. 10   d. 16

_____ 9. A(n) _____ is the lowest portion of the crankcase cavity in a four-stroke engine.

   a. engine sump   c. engine well

   b. oil tank   d. cleft

_____ 10. The _____ oil pump is most commonly used in full pressure lubrication systems.

   a. plunger   c. gear

   b. trochoid   d. centrifugal

Short Answers

11. What is oil viscosity and how is it determined?

12. What is the purpose of the oil relief valve and when does it operate?

13. What is the purpose of the oil filter bypass valve and when does it operate?

14. Describe the forced draft cooling system.

15. What is a thermostat and how is it used in a cooling system?

INSTRUCTOR VERIFICATION: _____

CHAPTER 8 Fuel Systems

Shop Assignment 8-1

Name: _____ Date: _____ Instructor: _____

Fuel Systems Knowledge Assessment

Objective

By correctly completing this assessment, you will demonstrate your knowledge of fuel system concepts and principles.

Directions

Answer the questions below using your textbook as a resource.

1. A 100-octane fuel will give your engine more power.

    True   **False**   *Circle the answer*

2. Detonation is dangerous to your engine.

    True   **False**   *Circle the answer*

3. High air temperature encourages detonation.

    True   **False**   *Circle the answer*

4. Technician A says that detonation is worse at higher altitude. Technician B says that a heavy load on the engine encourages detonation. Who is correct?

    A   **B**   *Circle the answer*

5. More combustion heat equals more engine power.

    True   **False**   *Circle the answer*

6. When speaking about air–fuel ratios, we are talking about pounds of fuel and pounds of air.

    True   **False**   *Circle the answer*

7. The primary purpose of a carburetor is to vaporize fuel.

   True   **False**   *Circle the answer*

8. If the fuel tank is higher than the carburetor, you need a fuel pump.

   True   **False**   *Circle the answer*

9. In California, the fuel tank must be vented to a charcoal canister.

   True   **False**   *Circle the answer*

10. A venturi has a modified hourglass shape.

    True   **False**   *Circle the answer*

11. Pressure rises in a venturi as the air velocity increases.

    True   **False**   *Circle the answer*

12. Technician A says the active carburetor circuit is identified by throttle position. Technician B says the active circuit is determined by engine rpm. Who is correct?

    A   **B**   *Circle the answer*

13. A vacuum fuel pump is also called an impulse pump.

    True   **False**   *Circle the answer*

14. The secondary idle ports give additional air routes for air and fuel at part throttle engine speeds.

    True ·   **False**   *Circle the answer*

15. Identify the types of carburetors by writing the type below the correct drawing.

    A   B   C

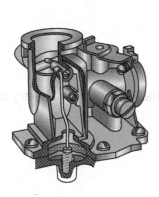
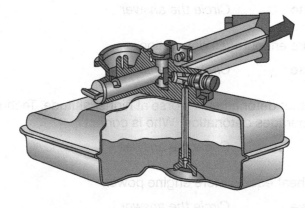

_____   _____   _____

INSTRUCTOR VERIFICATION:

Job Sheet 8-1

Name: _____ Date: _____ Instructor: _____

Vacuum Carburetor Inspection

Objective

After completing this job sheet, you should be able to correctly identify all parts and passages in a vacuum carburetor.

Directions

Use the instructions provided in the appropriate service manual to guide you through the inspection process.

Tools needed: Vacuum carburetor, service manual, contact or carb cleaner, eye protection

1. Identify the carburetor by noting the identification numbers on the side. Write the identification numbers here: _____

2. Remove the carb if it is installed on an engine.

3. Note the number of turns on the high- and low-speed adjustment screws. Write them here.

   High speed: _____

   Low speed: _____

4. According to the service manual, does this carb have the correct number of turns open for both adjustment screws?

   Yes   **No**   *Circle the answer*

5. Does this carb have a cold start device?

   Yes   **No**   *Circle the answer*

6. Note the type of cold start device.   **Bimetal   choke   plate   primer**   *Circle one*

7. Remove the high-speed screw and squirt the contact cleaner into the hole. Where does it come out?

   _____

8. Squirt the contact cleaner into the hole for the low-speed screw.

9. Where did the contact cleaner exit? _____

10. Replace the screws, remount the carb, and call your instructor over to check your work.

    _____

INSTRUCTOR VERIFICATION: _____

Job Sheet 8-1

Name: _____ Date: _____ Instructor: _____

Vacuum Carburetor Inspection

Objective

After completing this job sheet, you should be able to correctly identify all parts and passages in a vacuum carburetor.

Directions

Use the instructions provided. Use the appropriate service manual to guide you through the inspection process.

Tools needed: Vacuum carburetor, service manual, contact or carb cleaner, eye protection.

1. Identify the carburetor by noting the identification numbers on the side. Write the identification numbers here. _____

2. Remove the carb if it is installed on an engine.

3. Note the number of turns on the high- and low-speed adjustment screws. Write them here.

   High speed: _____

   Low speed: _____

4. According to the service manual, does this carb have the correct number of turns open for both adjustment screws?

   Yes   No   Circle the answer.

5. Does this carb have a cold start device?

   Yes   No   Circle the answer.

6. Note the type of cold start device.   Bimetal   choke   plate   primer   Circle one.

7. Remove the high-speed screw and squirt the contact cleaner into the hole. Where does it come out? _____

8. Squirt the contact cleaner into the hole for the low-speed screw.

9. Where did the contact cleaner exit? _____

10. Replace the screws, remount the carb, and call your instructor over to check your work.

INSTRUCTOR VERIFICATION: _____

Job Sheet 8-2

Name: _____ Date: _____ Instructor: _____

Float Carburetor Inspection

Objective
After completing this job sheet, you should be able to correctly identify the parts of a float carburetor, check the float level, and describe how the carburetor operates.

Directions
Use the instructions provided in the appropriate service manual to inspect this carb.

Tools needed: Float carburetor, service manual, float-level gauge, contact or carb cleaner, eye protection

1. Identify the carburetor by noting the identification numbers on the side. Write the identification numbers here: _____

2. Remove the float bowl and float along with the float valve.

3. Note the size of the main and pilot jets. Write them here.

   MJ: _____

   PJ: _____

4. According to the service manual, does this carb have the correct jets?

   Yes   **No**   *Circle the answer*

5. Check the float level according to instructions detailed in the manual and adjust if out of spec. Note: Some float-type carbs do not have adjustable floats.

6. Call the instructor over to check your float level.

7. Squirt the contact cleaner into the hole for the pilot jet. Where did the contact cleaner exit?

   _____

8. Replace all parts and call your instructor over for final verification.

   _____

INSTRUCTOR VERIFICATION:

Job Sheet 8-2

Name: _____ Date: _____ Instructor: _____

Float Carburetor Inspection

Objective

After completing this job sheet, you should be able to correctly identify the parts of a float carburetor, check the float level, and describe how the carburetor operates.

Directions

Use the instructions provided in the appropriate service manual to inspect this carb.

Tools needed: Float carburetor, service manual, float-level gauge, contact or carb cleaner, eye protection.

1. Identify the carburetor by noting the identification numbers on the side. Write the identification numbers here.
   _____

2. Remove the float bowl and float along with the float valve.

3. Note the size of the main and pilot jets. Write them here.
   MJ: _____
   PJ: _____

4. According to the service manual, does this carb have the correct jets?
   Yes   No   Circle the answer.

5. Check the float level according to instructions detailed in the manual and adjust it out of spec.
   Note: Some float-type carbs do not have adjustable floats.

6. Can the instructor over to check your float level.

7. Squirt the contact cleaner into the hole for the pilot jet. Where did this contact cleaner exit?
   _____

8. Replace all parts and call your instructor over for final verification.
   _____

Job Sheet 8-3

Name: _____ Date: _____ Instructor: _____

Diaphragm Carburetor Inspection

Objective

After completing this job sheet, you should be able to correctly identify the parts of a diaphragm carburetor and describe how the carburetor operates.

Directions

Use the instructions provided in the appropriate service manual to inspect this carb.

Tools needed: Diaphragm carburetor, service manual, contact or carb cleaner, eye protection

1. Identify the carburetor by noting the identification numbers on the side. Write the identification numbers here: _____

2. Remove the diaphragm, noting the air and fuel chambers. Is the air chamber vented to the atmosphere?

   Yes   **No**   *Circle the answer*

3. Note the number of turns on the high- and low-speed adjustment screws. Write them here.

   High speed: _____

   Low speed: _____

4. According to the service manual, does this carb have the correct number of turns open for both adjustment screws?

   Yes   **No**   *Circle the answer*

5. Does this carb have a cold start device?

   Yes   **No**   *Circle the answer*

6. Identify the type of cold start device;   **Bimetal   choke   plate   primer**   *Circle one*

7. Remove the high-speed screw and squirt the contact cleaner into the hole. Where does it come out?

   _____

8. Squirt the contact cleaner into the hole for the low-speed screw.

9. Where did the contact cleaner exit? _____

10. Replace the screws, remount the carb, and call your instructor over to check your work.

    _____

INSTRUCTOR VERIFICATION:

Job Sheet 8-3

Name: _____ Date: _____ Instructor: _____

Diaphragm Carburetor Inspection

Objective

After completing this job sheet, you should be able to correctly identify the parts of a diaphragm carburetor and describe how the carburetor operates.

Directions

Use the instructions provided in the appropriate service manual to inspect this carb.

Tools needed: Diaphragm carburetor, service manual, contact or carb cleaner, eye protection

1. Identify the carburetor by noting the identification numbers on the side. Write the identification numbers here: _____

2. Remove the diaphragm, noting the air and fuel chambers. Is the air chamber vented to the atmosphere?

   Yes   No   Circle the answer.

3. Note the number of turns on the high- and low-speed adjustment screws. Write them here.

   High speed: _____

   Low speed: _____

4. According to the service manual, does this carb have the correct number of turns open for both adjustment screws?

   Yes   No   Circle the answer.

5. Does this carb have a cold start device?

   Yes   No   Circle the answer.

6. Identify the type of cold start device: Bimetal   choke   plate   primer   Circle one.

7. Remove the high-speed screw and squirt the contact cleaner into the hole. Where does it come out?

   _____

8. Squirt the contact cleaner into the hole for the low-speed screw.

9. Where did the contact cleaner exit? _____

10. Replace the screws, remount the carb, and call your instructor over to check your work.

   _____

INSTRUCTOR VERIFICATION: _____

Job Sheet 8-4

Name: _____ Date: _____ Instructor: _____

Fuel Injection Component Location

Objective

After completing this job sheet, you should be able to correctly identify the parts of a fuel injection system and find their locations.

Directions

Use the appropriate service manual to locate the fuel injection sensors and components.

Tools needed: Fuel injected power equipment engine, service manual

1. Describe the location of the throttle body.

   _____

2. Describe the location of the fuel injector.

   _____

3. Describe the location of the MAP sensor.

   _____

4. Describe the location of the coolant temperature sensor.

   _____

5. Describe the location of the engine temperature sensor.

   _____

6. Describe the location of the ECU.

   _____

7. Describe the location of the cam position sensor.

   _____

8. Describe the location of the fuel pressure regulator.

   _____

9. Describe the location of the fuel filters.

    _____

10. Describe the location of the malfunction light.

    _____

INSTRUCTOR VERIFICATION: _____

CHAPTER 8

Test

True/False

Indicate whether the statement is true or false.

_____ 1. Higher octane fuel will make any engine produce more power.

_____ 2. High altitude discourages detonation.

_____ 3. Fuel valves are generally manual types.

_____ 4. Some air filters are made from foam.

_____ 5. A paper air filter must be oiled to work properly.

Multiple Choice

Identify the choice that best completes the statement or answers the question.

_____ 6. There are three types of fuel pumps: mechanical, vacuum, and _____.

    a. oscillating   c. electric

    b. reciprocating   d. static

_____ 7. A(n) _____ is used to regulate the flow of fuel from the fuel tank to the carburetor.

    a. choke plate   c. float valve

    b. accelerator pump d. secondary idle port

_____ 8. A(n) _____ makes cold starting much easier.

    a. choke plate   c. float valve

    b. accelerator pump   d. secondary idle port

_____ 9. A(n) _____ is used in part throttle operation.

    a. choke plate   c. float valve

    b. accelerator pump   d. secondary idle port

_____ 10. The main pickup tube of carburetor operation is used when the throttle valve is _____ open.

    a. 0–1/8   c. ¾–full

    b. ¼–¾   d. ¼–full

Short Answers

11. What does the venturi principle state and how does it apply to carburetor operation?

12. Define atomization. When does it take place in a carburetor?

13. Why are paper filters pleated and how do you service them?

14. Explain the need for correction inputs in a fuel injection system and note which sensors fulfill the role of correction inputs.

15. When would a vacuum fuel pump be used and how does it work?

INSTRUCTOR VERIFICATION: _____

CHAPTER 9

Throttle and Governor Control Systems

Shop Assignment 9-1

Name: _____ Date: _____ Instructor: _____

Governor and Throttle Control Knowledge Assessment

Objective
By correctly completing this assessment, you will demonstrate your knowledge of throttle and governor control systems.

Directions
Answer the questions below using your textbook as a resource.

1. Name the two types of governors. _____, _____.

2. Of the two types of governors, which is more responsive? _____

3. What is the purpose of a governor? _____

4. Name one advantage of an air vane over a centrifugal unit. _____

5. Name one advantage of a centrifugal unit over an air vane. _____

6. In a power equipment engine that uses a centrifugal governor, what position is the throttle valve in when starting the engine? _____

7. Name two other controls often linked to a manual throttle cable system: _____, _____.

8. Name three safeguards afforded by a governor control system: _____, _____, _____.

INSTRUCTOR VERIFICATION: _____

Throttle and Governor Control Systems

Shop Assignment 9-1

Name _____ Date: _____ Instructor _____

Governor and Throttle Control Knowledge Assessment

Objective

By correctly completing this assessment, you will demonstrate your knowledge of throttle and governor control systems.

Directions

Answer the questions below using your textbook as a resource.

1. Name the two types of governors. _____

2. Of the two types of governors, which is more responsive? _____

3. What is the purpose of a governor? _____

4. Name one advantage of an air vane over a centrifugal unit. _____

5. Name one advantage of a centrifugal unit over an air vane. _____

6. In a power equipment engine that uses a centrifugal governor, what position is the throttle valve in when starting the engine? _____

7. Name two other controls often linked to a manual throttle cable system. _____

8. Name three safeguards afforded by a governor control system. _____

INSTRUCTOR VERIFICATION: _____

Job Sheet 9-1

Name: _____ Date: _____ Instructor: _____

Air Vane Governor Inspection

Objective

After completing this job sheet, you will be able to correctly identify and set an air vane governor.

Directions

Use an appropriate service manual to perform the following air vane governor control inspection.

Tools

Tachometer, various hand tools

1. Remove the shroud covering the air vane assembly. Check to see that it moves freely.

2. Find the governor spring. What is it attached to? _____

3. Remove the link rod. Is it bent or distorted? _____

4. Remove the return spring. Does it hold the throttle wide open when the unit is at rest?
    _____

5. Some units have an anti-surge spring that keeps the engine from surging. Does your unit have
    this spring? _____

6. Reassemble the governor system.

7. Start the engine and note the engine speed with a tachometer.

8. What speed does the governor maintain? _____ rpm

9. If the governor spring can be adjusted, set your governor speed to 2,500 rpm.

10. Make the unit ready for inspection by the instructor.

INSTRUCTOR VERIFICATION: _____

Job Sheet 9-1

Name: _____   Date: _____   Instructor: _____

Air Vane Governor Inspection

Objective
After completing this job sheet, you will be able to correctly identify and set an air vane governor.

Directions
Use an appropriate service manual to perform the following air vane governor control inspection.

Tools
Tachometer, various hand tools

1. Remove the shroud covering the air vane assembly. Check to see that it moves freely.

2. Find the governor spring. What is it attached to? _____

3. Remove the link rod. Is it bent or distorted? _____

4. Remove the return spring. Does it hold the throttle wide open when the unit is at rest? _____

5. Some units have an anti-surge spring that keeps the engine from surging. Does your unit have this spring? _____

6. Reassemble the governor system.

7. Start the engine and note the engine speed with a tachometer.

8. What speed does the governor maintain? _____ rpm.

9. If the governor spring can be adjusted, set your governor speed to 2,500 rpm.

10. Make the unit ready for inspection by the instructor.

Job Sheet 9-2

Name: _____ Date: _____ Instructor: _____

Centrifugal Governor Inspection

Objective

After completing this job sheet, you will be able to correctly identify and set a centrifugal governor.

Directions

Use an appropriate service manual to perform the following centrifugal governor control inspection.

Tools

Tachometer, various hand tools

1. Loosen the governor arm from the governor shaft and remove it.

2. Remove the governor spring.

3. Remove the linkage rod. Is it bent or distorted? _____

4. Some units have a cable control to set governor speed. Is your unit equipped with a cable?
    _____

5. Some units have an anti-surge spring that keeps the engine from surging. Does your unit have this spring? _____

6. Reassemble the governor system.

7. Set the governor arm and the shaft it rides on. On most engines, the governor shaft is turned clockwise while the governor arm is held so that the throttle is closed.

8. Start the engine and note the engine speed with a tachometer.

9. What engine speed does the governor maintain? _____ rpm

10. If the governor spring can be adjusted, set your governor speed to 2,500 rpm.

11. Make the unit ready for inspection by the instructor.

INSTRUCTOR VERIFICATION: _____

Name: _____ Date: _____ Instructor: _____

Centrifugal Governor Inspection

Objective
After completing this job sheet, you will be able to correctly identify and set a centrifugal governor.

Directions
Use an appropriate service manual to perform the following centrifugal governor control inspection.

Tools
Tachometer, various hand tools

1. Loosen the governor arm from the governor shaft and remove it.

2. Remove the governor spring.

3. Remove the linkage rod. Is it bent or distorted? _____

4. Some units have a cable control to set governor speed. Is your unit equipped with a cable? _____

5. Some units have an anti-surge spring that keeps the engine from surging. Does your unit have this spring? _____

6. Reassemble the governor system.

7. Set the governor arm and the shaft it rides on. On most engines, the governor shaft is turned clockwise while the governor arm is held so that the throttle is closed.

8. Start the engine and note the engine speed with a tachometer.

9. What engine speed does the governor maintain? _____ rpm

10. If the governor spring can be adjusted, set your governor speed to 2,800 rpm.

11. Make the unit ready for inspection by the instructor.

INSTRUCTOR VERIFICATION _____

CHAPTER 9 Test

True/False

Indicate whether the statement is true or false.

_____ 1. Air vane style governors are less responsive than centrifugal types.

_____ 2. A centrifugal governor is less expensive to manufacture than an air vane governor.

_____ 3. Governors are needed to protect the engine from over-revving.

_____ 4. A pneumatic governor is the same as an air vane governor.

_____ 5. When centrifugal force is trying to close the throttle, it is opposed by spring tension.

Multiple Choice

Identify the choice that best completes the statement or answers the question.

_____ 6. A(n) _____ connects the governor arm to the throttle valve.

   a. air vane   c. throttle lever

   b. governor link   d. flyweights

_____ 7. The movement of the _____ is transferred outside the engine by the governor rod.

   a. governor spool   c. center ramp

   b. throttle valve   d. judder spring

_____ 8. The _____ opposes the governor weight.

    a. flyweight

    b. governor link

    c. governor spring

    d. governor spool

_____ 9. Increasing governor spring tension increases engine _____.

    a. compression

    b. backfiring

    c. governed speed

    d. life

_____ 10. A(n) _____ throttle control device gives the operator total control over the engine rpm.

    a. air vane governor

    b. centrifugal governor

    c. flyweight

    d. manual

Short Answers

11. Describe how the air vane governor works and its advantages over other types of governors.

12. Name the three main functions of any governor control system.

13. Explain how a centrifugal governor control system works and its advantage over the air vane governor.

14. Explain how a manual throttle control system works and mention the features normally found with this system.

15. Describe how the manual shut-off switch operates.

INSTRUCTOR VERIFICATION:

CHAPTER 10

Two-Stroke Engine Top End Inspection

Shop Assignment 10-1

Name: _____ Date: _____ Instructor: _____

Two-Stroke Top End Inspection Knowledge Assessment

Objective

By correctly completing this assessment, you should be able to demonstrate your knowledge of two-stroke engine inspection procedures.

Directions

Answer the questions below using your textbook as a resource.

1. What does a leak-down test tell you about engine condition?

   _____

2. How should the wrist pin clips be oriented during assembly?

   _____

3. How many places should you check the bore for out-of-round and taper? _____

4. Can a plated cylinder be rebored? _____

5. What is your best resource for the proper reassembly procedures for an engine?

   _____

6. What do small scratches on the piston skirt indicate?

   _____

7. Name two ways to clean the cylinder after it is rebored.

    _____

8. How do you dress small scratches on the piston skirt?

    _____

9. What does a small arrow stamped into the piston crown mean?

    _____

10. What would happen if you installed the piston backwards?

    _____

11. Is break-in important? _____

12. Can old piston rings be reused if they look fine? _____

13. How do you deglaze a cylinder? _____

INSTRUCTOR VERIFICATION: _____

Job Sheet 10-1

Name: _____ Date: _____ Instructor: _____

Two-Stroke Top End Inspection

Objective

After completing this job sheet, you should be able to correctly inspect the piston, cylinder, and reed valve of a two-stroke engine.

Directions

Using a two-stroke engine designated for this job sheet and the appropriate service manual, remove the head, cylinder, and piston. Measure the piston-to-cylinder clearance and correctly reassemble.

Tools Needed

Assorted hand tools, micrometer, dial bore gauge, service manual, feeler gauges, wrist pin remover

1. Loosen the cylinder head bolts using a crisscross pattern.

2. Remove the cylinder and its reed valve assembly.

3. Remove the wrist pin clip and slide the wrist pin out of the piston. *Note*: On running engines, you may have to use a wrist pin remover.

4. Check the ring groove clearance with a feeler gauge.

   Manual specification for ring groove clearance: _____ mm

   Measured clearance: _____ mm

5. Remove the rings and square one up in the cylinder using the piston crown. Check the end gap.

   Manual specification for ring end gap: _____ mm

   Measured clearance: _____ mm

6. Check the bore size in six places. Write the bore sizes in the table below.

x-axis	*y*-axis	Out-of-Round
Top		
Middle		
Bottom		
Taper	Taper	

7. Does this cylinder meet service manual specifications for taper and out-of-round?

   Yes   **No**   *Circle the answer*

8. Measure the piston size: _____ mm

9. Calculate the piston-to-cylinder clearance:

   Top: _____ mm
   Middle: _____ mm
   Bottom: _____ mm

10. Does this engine meet spec. for clearance? Use the largest number you calculated.

    Yes   **No**   *Circle the answer*

11. Remove the reed assembly. Do the reeds touch the reed cage?

    Yes   **No**   *Circle the answer*

INSTRUCTOR VERIFICATION: _____

CHAPTER 10

Test

True/False

Indicate whether the statement is true or false.

_____ 1. You must remove the cylinder head to service the top end of a two-stroke engine.

_____ 2. A dial bore gauge can be used to check piston-to-cylinder clearance.

_____ 3. A dial bore gauge will tell you how much taper there is in the cylinder.

_____ 4. "Scuffing" and "scoring" are two words for the same thing.

_____ 5. A stuck piston ring often results from overheating.

Multiple Choice

Identify the choice that best completes the statement or answers the question.

_____ 6. _____ is the difference between the largest cylinder bore measurement and the smallest on the same axis.

    a. Out-of-round   c. Taper

    b. Parallelism   d. Warpage

_____ 7. The side of the piston running at 90° to the length of the wrist pin axis is called the _____.

    a. thrust face   c. ring land

    b. crown   d. pin boss

_____ 8. The _____ is the uncut area between the ring grooves.

    a. thrust face   c. ring land

    b. crown   d. boss

_____ 9. A _____ is used to measure piston ring end gap.

    a. micrometer   c. caliper

    b. feeler gauge   d. dial gauge

_____ 10. Deglazing is done correctly if the crosshatch angle is _____.

    a. 45°   c. 90°

    b. 60°   d. 120°

Short Answers

11. Explain how you perform a compression check and what it is actually measuring.

12. Explain the importance of proper piston-to-cylinder clearance.

13. Explain why pistons are egg-shaped and how you measure a piston's diameter.

14. Describe how a technician would diagnose a failed air cleaning system.

15. What procedure should you use to start a freshly rebuilt engine?

INSTRUCTOR VERIFICATION:

CHAPTER 11

Two-Stroke Engine Bottom End Inspection

Shop Assignment 11-1

Name: _____ Date: _____ Instructor: _____

Two-Stroke Bottom End Inspection Knowledge Assessment

Objective
By correctly completing this assessment, you should be able to demonstrate your knowledge of two-stroke engine bottom end inspection procedures.

Directions
Answer the questions below using your textbook as a resource.

1. Crank seals should be replaced every time the crankshaft bearings are replaced.

   True　**False**　*Circle the answer*

2. A leaky crank seal can result in a lean running condition.

   True　**False**　*Circle the answer*

3. You must remove the top end first to service the crankshaft.

   True　**False**　*Circle the answer*

4. Heating the crankcases will allow for easier disassembly.

   True　**False**　*Circle the answer*

5. The rubber lip of an engine seal must be "live."

   True　**False**　*Circle the answer*

6. Freezing the bearing and heating engine case will ease bearing installation.

   True   **False**   *Circle the answer*

7. Use a dial bore gauge to measure connecting rod side clearance.

   True   **False**   *Circle the answer*

8. Install thrust washers with the chamfered side facing the thrust load.

   True   **False**   *Circle the answer*

9. Some engines require a case splitter tool when splitting the cases.

   True   **False**   *Circle the answer*

10. The crankshaft from a two-stroke engine is usually a single-piece design.

    True   **False**   *Circle the answer*

INSTRUCTOR VERIFICATION:

Shop Assignment 11-2

Name: _____ Date: _____ Instructor: _____

Two-Stroke Failure Modes and Symptoms

Objective

By correctly completing this worksheet, you should be able to demonstrate your knowledge of common two-stroke engine failure modes and their symptoms.

Directions

Fill in the blank cells below using your textbook as a resource.

Symptom	Cause	Remedy
Growling noise from engine while running		
Engine will not pass leak-down test		
Knocking noise from engine		
Engine starts but won't turn freely		
Crankshaft will not turn		

INSTRUCTOR VERIFICATION: _____

Shop Assignment 11-2

Name: _____ Date: _____ Instructor: _____

Two-Stroke Failure Modes and Symptoms

Objective

By correctly completing this worksheet, you should be able to demonstrate your knowledge of common two-stroke engine failure modes and their symptoms.

Directions

Fill in the blanks below using your textbook as a reference.

Symptom	Cause	Remedy
Growling noise from engine while running		
Engine will not pass lock-down test		
Knocking noise from engine		
Engine starts but won't turn freely		
Crankshaft will not turn		

INSTRUCTOR VERIFICATION

Job Sheet 11-1

Name: _____ Date: _____ Instructor: _____

Two-Stroke Bottom End Inspection

Objective

After completing this job sheet, you should be able to correctly remove and replace the crankshaft from a modern two-stroke engine and check it for radial play.

Directions

Using a training aid designated for this job sheet, split the cases, remove the crankshaft, check it for radial play with V blocks and side clearance with the appropriate feeler gauge, then replace the crankshaft, and mate the case halves together. *Note*: Some two-stroke engines are designed to be split from one side. Failure to split the cases from the correct side will dramatically increase the difficulty of this project. Check the service manual for directions.

Tools Needed

Assorted hand tools, dead blow hammer, V blocks, a set of feeler gauges, set of engine cases with no top end installed

1. Double check that all case screws or bolts are removed before attempting to split the cases.

2. Before attempting to split the case, look for pry points. These are areas used to pry the cases apart without damaging the mating surfaces. A few taps with a dead blow hammer will part the cases once you have them pried apart.

3. Remove the crankshaft from the opposite case. Try tapping it lightly with a dead blow hammer to see if it should slide out of its main bearing. If not, you may need to place the case half in a hydraulic press or use a crank removal tool to remove it from the case.

INSTRUCTOR VERIFICATION: _____

4. Support the ends of the crankshaft with V blocks. Spin the shaft to make sure it was not knocked out of alignment during the removal process. The ends of the crankshaft should not wobble up and down as you spin the crank. If they do, the crank is out of true. The crank should also spin freely on the V blocks. If it does not, the big end bearing is damaged.

5. Use feeler gauges to measure the clearance between the big end of the connecting rod and the crank. Consult the service manual for the proper clearance. Excessive clearance indicates worn thrust washers.

   Service manual spec. for big end side clearance: _____ mm

   Measured clearance: _____ mm

6. Hold the crankshaft in one hand and gently move the connecting rod up and down with the other. There should be no detectable play.

INSTRUCTOR VERIFICATION: _____

7. Pull the crankshaft through its main bearing. Usually, a tap with a dead blow hammer will seat the crank against its bearing. Install the remaining crankcase half. Pay special attention to the mating surface. If a gasket is not used, apply liquid gasket sealer to ensure an airtight fit.

INSTRUCTOR VERIFICATION: _____

8. Tap the case lightly with a dead blow hammer until the mating surfaces make contact.

9. Reinstall the case bolts and tighten to the specified torque.

10. Make the engine ready for final inspection.

INSTRUCTOR VERIFICATION: _____

CHAPTER
11

Test

True/False

Indicate whether the statement is true or false.

_____ 1. Most two-stroke engines require you to remove the top end to split the engine cases.

_____ 2. An engine seal can be reused after removal if it is "live."

_____ 3. Rubber engine seals can be reused.

_____ 4. Put a ball bearing in the freezer before installation.

_____ 5. Ball bearings are held in the engine cases by a press fit.

Multiple Choice

Identify the choice that best completes the statement or answers the question.

_____ 6. The _____ is the most commonly replaced part of a two-stroke crankshaft.
   a. thrust washer   c. connecting rod
   b. wrist pin   d. main bearing

_____ 7. _____ is required to ensure that there is enough space between the rod and flywheels.
   a. Side clearance   c. Run out
   b. Radial clearance   d. Bearing float

_____ 8. _____ is the up-and-down motion of the connecting rod at the lower bearing.

   a. Side clearance   c. Run out

   b. Radial clearance   d. Bearing float

_____ 9. _____ is checked with dial indicators to ensure trueness of the crankshaft.

   a. Side clearance   c. Run out

   b. Radial clearance   d. Bearing float

_____ 10. A _____ is used to pinpoint the location of a bad bearing.

   a. Dial indicator   c. stethoscope

   b. snap gauge   d. feeler gauge

Short Answers

11. Explain the procedure for inspecting a crankshaft and the three measurements a technician should make.

12. Describe the procedure for doing a thorough bench test on a freshly reassembled two-stroke engine. Why this is important?

13. Describe the procedure for checking a two-stroke engine for bad main bearings. What tool is useful in pinpointing the bad bearing?

14. Describe the procedure for inspecting the engine crankcases. Why it is important that it be thorough?

15. Describe the importance of the crank seals. How should they be treated upon reassembly?

INSTRUCTOR VERIFICATION:

Four-Stroke Engine Inspection

Shop Assignment 12-1

Name: _____ Date: _____ Instructor: _____

Four-Stroke Top End Inspection Knowledge Assessment

Objective
By correctly completing this assessment, you should be able to demonstrate your knowledge of four-stroke engine top end inspection procedures.

Directions
Answer the questions below using your textbook as a resource.

1. Name two tests that should always be performed before a four-stroke engine is disassembled.

   _____

2. What does air escaping from the exhaust pipe during a leak-down test tell the technician?

   _____

3. Name a common tool for cleaning piston ring grooves.

   _____

4. Can you use an old piston if it is still in spec.?

   _____

5. How do you know if you installed the piston rings correctly?

   _____

6. If the spark plug threads are ruined, must you replace the head?

   _____

7. If you find that the engine you are rebuilding has a warped head, what should you do?

_____

8. How should you clean valves before inspecting them?

_____

9. How should valve springs be installed in an engine?

_____

10. Should you reface stellite-coated valves?

_____

11. How do you prepare new guides after installation?

_____

12. Should you cut the valve seats if you replaced the valve guide?

_____

13. Should you reuse old engine seals?

_____

14. You are lapping in valves after the seats have been cut, but you notice that one of the valves is not hitting all the way around the face of the valve. What does this tell you?

_____

15. Why should you break in a newly rebuilt engine?

_____

16. Where should the piston be when timing a camshaft on most engines?

_____

17. What is the correct temperature for adjusting valves?

_____

18. How can you tell if the camshaft is correctly timed?

_____

19. Name the three angles most modern valves have.

_____

INSTRUCTOR VERIFICATION: _____

Job Sheet 12-1

Name: _____ Date: _____ Instructor: _____

Four-Stroke Top End Inspection

Objective
After completing this job sheet, you should be able to disassemble, inspect, and time the camshaft on a single-cylinder four-stroke engine.

Directions
Using a training aid designated for this job sheet, remove the camshaft, cylinder head, and cylinder; then, reassemble the top end while correctly timing the camshaft.

Tools
Various hand tools

1. Obtain a training aid designated for this project. Use the service manual as your resource for step-by-step instructions.

2. Set the crankshaft to top-dead center on the compression stroke (TDCC). Remove the top cover.

3. Note the camshaft timing marks. Are they lined up properly?
   Yes   **No**   *Circle the answer*

4. Remove the camshaft.

5. Remove the cylinder.

6. Remove the wrist pin and then the piston.

INSTRUCTOR VERIFICATION:

7. Measure the piston diameter and the cylinder bore diameter. Calculate the piston-to-cylinder clearance. _____ mm

8. What is the factory spec for piston-to-cylinder clearance? _____ mm

9. Is the cylinder clearance in spec.?
   Yes   **No**   *Circle the answer*

10. Measure the cylinder at the top, middle, and bottom on the *x* axis. Record your measurements in the table below, then measure the top, middle, and bottom of the cylinder on the *y* axis and record below. You will now calculate the taper and out-of-round.

	x axis	*y* axis	Out-of-Round
Top			
Middle			
Bottom			
	Taper	Taper	

11. Install the piston. Use the old wrist pin clips for this exercise if you are using a non-running engine. If this is a running engine use new wrist pin clips.

12. Remove the valves and examine the valve seats by using Prussian blue. Are the seats located correctly on the face of the valves?

    Yes   **No**   *Circle the answer*

13. Replace the valves. (You may need a valve spring compressor for this operation.)

14. Carefully install the piston into the cylinder. Make sure the piston ring end gaps are properly staggered and that you do not snag any rings on the cylinder skirt. You may elect to use a piston ring compressor.

INSTRUCTOR VERIFICATION: _____

15. Install the camshaft. Make sure the crankshaft is at top-dead center (TDC) and that the "T" is adjacent to the index mark.

INSTRUCTOR VERIFICATION: _____

16. Replace the valve cover and measure the valve lash. Some engines use valve stem caps to adjust valve lash, which are ground on a valve-grinding machine while others use a screw-and-locknut arrangement. Factory spec. for valve lash:

    Intake: _____mm

    Exhaust: _____mm

17. Make the engine ready for final inspection.

INSTRUCTOR VERIFICATION: _____

CHAPTER 12 — Test

True/False

Indicate whether the statement is true or false.

_____ 1. A four-stroke engine requires you to remove it from the implement it is attached to before removing the top end.

_____ 2. The middle ring on a piston scrapes the excess oil from the cylinder wall when the piston is moving downward.

_____ 3. Preignition occurs when the fuel–air mixture ignites before the spark plug fires.

_____ 4. The most common cause of detonation is a spark plug with the wrong heat range.

_____ 5. A compression check is the most comprehensive diagnostic test you can make on an engine.

Multiple Choice

Identify the choice that best completes the statement or answers the question.

_____ 6. The _____ ring is designed to seal most of the combustion pressure on top of the piston.

    a. compression   c. oil scraper

    b. oil control   d. O

_____ 7. The _____ ring is designed to remove most of the oil from the cylinder wall.

    a. compression   c. oil scraper

    b. oil control   d. O

_____   8. Air leaking from the intake manifold during a leak-down test means the _____ is/are leaking.

    a. exhaust valves

    b. intake valves

    c. piston rings

    d. head gasket

_____   9. Piston ring gaps should be _____ apart on the piston during installation.

    a. 90°

    b. 180°

    c. 60°

    d. 120°

_____   10. Use a _____ to remove the valves from the head for inspection.

    a. dead blow hammer

    b. case splitter

    c. spring compressor

    d. vernier caliper

Short Answers

11. Describe the procedure for inspecting a four-stroke engine's valves.

12. How do you check a valve guide and what tools will you need?

13. Explain how a technician diagnoses engine problems.

14. How would you replace valve guides and what precautions should you observe?

15. Describe how you lap valve seats.

INSTRUCTOR VERIFICATION:

CHAPTER 13

Fundamentals of Electricity

Shop Assignment 13-1

Name: _____ Date: _____ Instructor: _____

Electrical Principles Knowledge Assessment

Objective
By correctly completing this assessment, you should be able to demonstrate your knowledge of electrical fundamentals.

Directions
Answer the questions below using your textbook as a resource.

1. An electrical component that has an anode and a cathode is called a _____.

2. Do you hook your ammeter up in **series** or **parallel** to measure amperage? *Circle the answer.*

3. To measure battery voltage, you would set your meter to _____ volt(s).

4. Where would you find AC voltage?

   _____

5. Calculate the resistance in a 12-volt circuit that has 2 amperes of current. _____

6. What would be an acceptable voltage drop across an ignition switch? _____
   volt(s) DC.

7. How much voltage should there be after the load? _____

8. Most manufacturers generally use the **conventional** or **electron** theory. *Circle the answer.*

9. A transistor is an example of a(n) _____.

10. Name a type of semiconductor found in power equipment. _____

11. _____ is a measure of electrical pressure.

12. _____ is a measure of current flow.

13. _____ is a measurement of opposition to current flow.

14. What is the unit of measurement for resistance? _____

15. Is house current **AC** or **DC**? *Circle the answer.*

16. Since an entire schematic can cover several pages, what do you call a portion of the schematic that deals only with the circuit you are interested in?

   _____

17. Are color codes consistent among all manufacturers?

   Yes   **No**   *Circle the answer*

18. A customer comes into your shop and complains that his battery goes dead. Which system is probably at fault?

   Ignition   **Charging**   **Starting**   **Lighting**   *Circle the answer*

19. An unwanted path to ground before the load is called a(n) _____.

20. What setting should your meter be at for measuring charging system current?

   _____

21. What setting should your meter be at for measuring voltage drop across a switch?

   _____

22. What setting should your meter be at for measuring charging voltage at the battery?

   _____

23. What would happen if you set your meter to DC volts and tried to measure charging current?

   _____

24. A circuit with more than one path to ground is called a(n) _____.

INSTRUCTOR VERIFICATION: _____

Shop Assignment 13-2

Name: _____ Date: _____ Instructor: _____

Reading Schematic Diagrams

Objective

After completing this shop assignment, you should be able to read a simple schematic and create a block diagram.

Directions

Answer the questions below using the schematic picture.

1. What color is the ground wire in the schematic shown on the following page? _____

2. Does this vehicle have to be in neutral to start?

   Yes   **No**   *Circle the answer*

3. Does the clutch switch have to be closed for the starter motor to turn over?

   Yes   **No**   *Circle the answer*

4. What color wire supplies voltage to the start switch? _____

5. From where does the start switch get its power? _____

6. Which wires have continuity (are connected to each other) when the ignition switch is in the Off position? _____

7. Now, using the schematic on the next page, make a block diagram of the starting system. Your diagram should include the battery, starter motor, starter relay, neutral switch, ignition switch, starter switch, and clutch switch. Use a separate piece of paper to practice making your block diagram, then use the box provided for your final version.

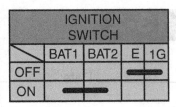

ENGINE STOP SWITCH		
	E	IG
OFF	▬▬	▬
RUN		

STARTER SWITCH		
	ST	BAT3
FREE		
PUSH	▬▬	▬

IGNITION SWITCH				
	BAT1	BAT2	E	1G
OFF			▬▬	▬
ON	▬▬	▬		

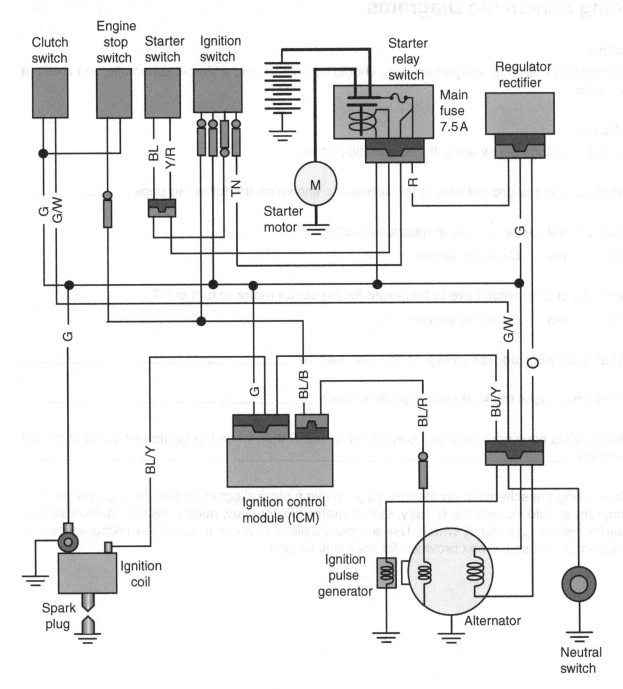

INSTRUCTOR VERIFICATION: _____

Job Sheet 13-1

Name: _____ Date: _____ Instructor: _____

Meter Usage

Objective
After completing this job sheet, you should be able to use your meter to measure volts, amperage, and ohms.

Directions
Use your multi-meter, the training aid designated for this job sheet, and the appropriate service manual to answer the questions below.

Tools Needed
Multi-meter; power equipment with battery-powered circuits, such as a riding lawnmower

1. Locate the battery and place the black meter lead of your meter on the battery's negative battery post. Place the red meter lead on the positive post. Set the meter to "DCV." What reading do you get? _____ DCV

2. What kind of connection did you just make?

   Series   **parallel**   *Circle the answer*

3. Set your meter to AC volts and select the "200 volt" scale. Plug the meter leads into a wall socket. How much AC voltage did you measure? _____

4. Remove the headlight from a garden tractor or equivalent and disconnect the wire from the back of the headlight. Now consult the manual and determine which of the wires is the ground wire. Set your meter to "Ohms" and place the black meter lead on the ground wire terminal and the red meter on any bare metal on the chassis. What reading did you get? _____ ohms

5. Disconnect the negative battery cable. Set your meter to "DCA." Move the meter leads to the correct jacks on the meter for measuring amperage. Hook your meter's black lead to the negative battery post. Hook your red meter lead to the negative battery cable. Turn the ignition switch on. Do not start the vehicle! What reading did you get? _____ DCA

6. Set your meter to "Ohms" and move the meter leads to the proper jacks for measuring resistance. Place one meter lead on one terminal of the headlight you removed in Step 4. Place the other meter lead on the ground terminal. What reading did you get? _____

INSTRUCTOR VERIFICATION: _____

Name: _____   Date: _____   Instructor: _____

Meter Usage

Objective

After completing this job sheet, you should be able to use your meter to measure volts, amperage, and ohms.

Directions

Use your multimeter, the training aid designated for this job sheet, and the appropriate service manual to answer the questions below.

Tools Needed

Multimeter, power equipment with battery-powered circuits, such as a riding lawnmower.

1. Locate the battery and place the black meter lead of your meter on the battery's negative battery post. Place the red meter lead on the positive post. Set the meter to "DCV." What reading did you get? _____ DCV.

2. What kind of connection did you just make?

   Series   Parallel   Circle the answer.

3. Set your meter to AC volts and select the "300 volt" scale. Plug the meter leads into a wall socket. How much AC voltage did you measure? _____

4. Remove the headlight from a garden tractor or equivalent and disconnect the wire from the back of the headlight. Now consult the manual and determine which of the wires is the ground wire. Set your meter to "Ohms" and place the black meter lead on the ground wire terminal and the red meter on any bare metal on the chassis. What reading did you get? _____ ohms.

5. Disconnect the negative battery cable. Set your meter to "DCA." Move the meter leads to the correct jacks on the meter for measuring amperage. Hook your meter's black lead to the negative battery post. Hook your red meter lead to the negative battery cable. Turn the ignition switch on. Do not start the vehicle! What reading did you get? _____ DCA.

6. Set your meter to "Ohms" and move the meter leads to the proper jacks for measuring resistance. Place one meter lead on one terminal of the headlight you removed in Step 4. Place the other meter lead on the ground terminal. What reading did you get? _____

INSTRUCTOR VERIFICATION: _____

Job Sheet 13-2

Name: _____ Date: _____ Instructor: _____

Voltage Drop Testing

Objective
After completing this job sheet, you should be able to do simple voltage drop testing.

Directions
Using a training aid designated for this job sheet, follow the instructions and answer the questions.

1. Set your meter to "DCV." Move the meter leads to the correct meter jacks for measuring DC voltage. Place the red meter lead on one side of the main fuse. (Do not remove the fuse.) Place the other meter lead on the opposite side of the main fuse. Turn the ignition switch on. What reading did you get? _____ DCV

2. You just did a voltage drop test. You measured how much voltage was dropped across the main fuse.

3. Find a connector in the wiring harness. The connector must be part of an active circuit when the ignition switch is turned on. Place your meter leads across the connector and turn the ignition switch on. What reading did you get? _____ DCV

INSTRUCTOR VERIFICATION: _____

Name _____ Date: _____ Instructor _____

Voltage Drop Testing

Objective

After completing this job sheet, you should be able to do simple voltage drop testing.

Directions

Using a training aid designated for this job sheet, follow the instructions and answer the questions.

1. Set your meter to "DCV." Move the meter leads to the correct meter jacks for measuring DC voltage. Place the red meter lead on one side of the main fuse. (Do not remove the fuse.) Place the other meter lead on the opposite side of the main fuse. Turn the ignition switch on. What reading did you get? _____ DCV.

2. You just did a voltage drop test. You measured how much voltage was dropped across the main fuse.

3. Find a connector in the wiring harness. The connector must be part of an active circuit when the ignition switch is turned on. Place your meter leads across the connector and turn the ignition switch on. What reading did you get? _____ DCV.

INSTRUCTOR VERIFICATION.

CHAPTER 13

Test

True/False

Indicate whether the statement is true or false.

_____ 1. The conventional theory of electricity states that current flows from the positive to the negative terminal.

_____ 2. Good conductors have electrons that are difficult to knock out of their orbits.

_____ 3. Electricity can produce magnetism and magnetism can produce electricity.

_____ 4. A capacitor can store electricity.

_____ 5. The flow of protons through a circuit is called electrical current.

Multiple Choice

Identify the choice that best completes the statement or answers the question.

_____ 6. A(n) _____ circuit has one path to ground.

    a. parallel   c. series/parallel

    b. open   d. series

_____ 7. A(n) _____ circuit has more than one path to ground.

    a. parallel   c. series/parallel

    b. open   d. series

_____ 8. A(n) _____ circuit has an incomplete path for current flow.

    a. parallel   c. series/parallel

    b. open   d. series

_____ 9. A(n) _____ circuit contains a load in series and a parallel load in the same circuit.

    a. parallel   c. series/parallel

    b. open   d. series

_____ 10. A(n) _____ circuit has a path to the source of power before it reaches the load.

    a. short   c. Grounded

    b. parallel   d. Incomplete

Short Answers

11. Using the analogy of water, describe how voltage flows and how resistance acts on that flow.

12. Define current and name the units with which it is measured.

13. What is electrical potential and what units do we use to measure it?

14. Describe the relationship among resistance, voltage, and current flow and state the law that governs their relationship.

15. Explain the difference between direct current and alternating current, and give at least one example of where you might find each in a power equipment electrical system.

INSTRUCTOR VERIFICATION:

CHAPTER 14

Power Equipment Engine Charging Systems and DC Circuits

Shop Assignment 14-1

Name: _____ Date: _____ Instructor: _____

Charging Systems Knowledge Assessment

Objective
By correctly completing this assessment, you should be able to demonstrate your knowledge of charging system fundamentals.

Directions
Answer the questions below using your textbook as a resource.

1. All charging systems function in basically the same way.

   True   False   *Circle the answer*

2. Technician A says that an automotive charger works fine for charging power equipment batteries as long as you set it to a low charge level. Technician B says that a good power equipment charger should be of a constant current type. Who is correct?

   A   B   *Circle the answer*

3. A half-wave charging system uses a single diode to rectify the current.

   True   False   *Circle the answer*

4. How is a three-phase excited-field charging system regulated?

   a. By grounding one or more stator leads

   b. By looping excess current back to the stator coil

   c. By reducing the current flowing through the field coil

   d. By shunting excess current to ground

5. A battery can be recharged by either AC or DC depending on the type of battery.

   True   False   *Circle the answer*

6. A customer complains that his battery goes dead while sitting. Which components can be eliminated as faulty?   *Circle all that apply*

   a. The stator coil

   b. The rotor

   c. The rectifier

   d. The wiring

7. You start a lawn tractor and connect an ammeter to the battery. The meter reads +2 DCA. Is this system charging?

   Yes   No   *Circle the answer*

8. A customer complains that her battery goes dead when her tractor sits for a day. You connect your ammeter to the battery and note that with the key off the meter reads 2 DCA. Is this normal current drain?

   Yes   No   *Circle the answer*

9. Technician A says that you must use a known good battery to check a charging system. Technician B says that you should use the battery that came with the vehicle. Who is correct?

   A   B   *Circle the answer*

10. How do you check for a bad field coil?

    a. Do a resistance check and compare your readings to the factory spec.

    b. Field coils cannot go bad; they are solid state.

    c. Measure the current draw across the slip rings.

    d. Check the AC output and compare to the factory spec.

11. To check a stator coil for AC output, you unplug it from the charging system and connect your meter leads to each side of the stator. What is the next step?

    a. Start the engine and observe the AC output.

    b. Set your meter to "Ohms" and check its resistance.

    c. Measure the voltage drop across the coil.

    d. Never check an unplugged stator; it may burn out.

12. What would be a typical resistance for each leg of a stator coil?

    a. 1–10 ohms

    b. 2–100 ohms

    c. 0.2–1 ohm

    d. infinity

13. What would be the typical charge voltage when checked at the battery for a properly operating charging system?

    a. 8–12 DDC

    b. 10–20 VAC

    c. 13–15 VDC

    d. 1–2 ohms

14. How is a three-phase permanent-magnet system regulated?

    a. By reducing current to the field coil

    b. By shunting one or more of the stator leads to ground

    c. By disconnecting one or more stator leads from the rectifier

    d. By grounding the rectifier leads

15. A technician wants to test a three-phase excited-field charging system; so he connects his meter in series to the battery and sets his meter to "Amps." He cranks the engine until it starts. What did he do wrong?

    a. He started the machine with his meter in the circuit.

    b. He should have put his meter in parallel for an amp check.

    c. He did not let the machine warm up.

    d. He forgot to turn on all the lights.

INSTRUCTOR VERIFICATION:

13. What would be the typical charge voltage when checked at the battery for a properly operating charging system?

   a. 8–12 DDC

   b. 40–60 VAC

   c. 13–15 VDC

   d. 4–6 ohms

14. How is a three-phase permanent-magnet system regulated?

   a. By reducing current to the field coil.

   b. By shunting one or more of the stator leads to ground.

   c. By disconnecting one or more stator leads from the rectifier.

   d. By grounding the rectifier leads.

15. A technician wants to test a three-phase excited-field charging system, so he connects his meter in series to the battery and sets his meter to "Amps." He cranks the engine until it starts. What did he do wrong?

   a. He started the machine with his meter in the circuit.

   b. He should have put his meter in parallel for an amp check.

   c. He did not let the machine warm up.

   d. He forgot to turn on all the lights.

INSTRUCTOR VERIFICATION:

Job Sheet 14-1

Name: _____ Date: _____ Instructor: _____

Charging System Inspection #1

Objective
After completing this job sheet, you should be able to accurately evaluate any charging system.

Directions
Check at least one example of each type of charging system available. Since some tests listed on the job sheet will not apply to all charging systems, mark "NA" on tests that do not apply. Use your multi-meter, a training aid designated for this job sheet, and the appropriate service manual to answer the questions below.

Tools Needed
Multi-meter

1. Identify the type of charging system by the number of leads on the stator.

    a. **1,** half wave

    b. **2,** full wave   *Circle one*

    c. **3,** three phase

2. Is this a **permanent-magnet** or an **excited-field** charging system? *Circle one*

3. **Key-off amp draw test.** Determine if the electrical system is draining the battery when the key is switched off by hooking your meter in series to the negative side of the battery and observing any current draw with the key off. Note that some circuits with clocks or computer memory should draw 0.1–0.2 milliamps with the key off.

    Current draw: _____

4. **Charging voltage test.** Check the charging voltage at the battery. Hook your meter in parallel to the battery and let the vehicle idle.

    Voltage at idle: _____

5. **Charging current test.** Remove the main fuse and hook your meter in series at the fuse holder after you have configured the meter to test DCA. You should now be able to start the machine with the meter hooked up in series without damage to it. Observe the charging current at idle.

    Current at idle: _____

6. **Stator output test.** Unplug all the stator leads and hook the meter to one pair with the meter set at ACV. Test all three pairs of leads. Start the machine and record the voltage at idle.

   ACV at idle: 1–2 _____ ACV

   2–3 _____ ACV

   1–3 _____ ACV

7. **Field coil test.** If the system you are testing has an excited-field rotor, you may be able to check the field coil and brushes. Refer to the manual for details on testing the field coil.

   a. What is the manual spec for brush length? _____ mm

   b. What is the manual spec for field coil resistance? _____ ohms

8. **Rectifier test.** Unplug the regulator/rectifier and check the forward and reverse bias of the diodes in the rectifier. Note that you should have two checks for testing a half-wave system, four checks for a full-wave system, and six checks for a three-phase system.

   a. forward bias: _____

   _____

   _____

   b. reverse bias: _____

   _____

   _____

9. Is this charging system working correctly?

   Yes   No   *Circle one*

10. If you circled "No," explain what is wrong with the system.

   _____

   _____

   _____

INSTRUCTOR VERIFICATION: _____

Job Sheet 14-2

Name: _____ Date: _____ Instructor: _____

Charging System Inspection #2

Objective

After completing this job sheet, you should be able to accurately evaluate the charging system on any type of power equipment.

Directions

Check at least one example of each type of charging system available. Since some tests listed on the job sheet will not apply to all charging systems, mark "NA" on tests that do not apply. Use your multi-meter, a training aid designated for this job sheet, and the appropriate service manual to answer the questions below.

Tools Needed

Multi-meter

1. Identify the type of charging system by the number of leads on the stator.

    a. **1**, half wave

    b. **2**, full wave   *Circle one*

    c. **3**, three phase

2. Is this a **permanent-magnet** or an **excited-field** charging system? *Circle one*

3. **Key-off amp draw test.** Determine if the electrical system is draining the battery when the key is switched off by hooking your meter in series to the negative side of the battery and observing any current draw with the key off. Note that some systems with clocks or computer memory should draw 1–2 milliamps with the key off.

    Current draw: _____

4. **Charging voltage test.** Check the charging voltage at the battery. Hook your meter in parallel to the battery and let the vehicle idle.

    Voltage at idle: _____

5. **Charging current test.** Remove the main fuse and hook your meter in series at the fuse holder after you have configured the meter to test DCA. You should now be able to start the machine with the meter hooked up in series without damage to it. Observe the charging current at idle.

    Current at idle: _____

6. **Stator output test.** Unplug all the stator leads and hook the meter to one pair with the meter set at ACV. Test all three pairs of leads. Start the machine and record the voltage at idle.

   ACV at idle: 1–2 _____ ACV

   2–3 _____ ACV

   1–3 _____ ACV

7. **Field coil test.** If the system you are testing has an excited-field rotor, you may be able to check the field coil and brushes. Refer to the manual for details on testing the field coil.
   a. What is the manual spec for brush length? _____ mm
   b. What is the manual spec for field coil resistance? _____ ohms

8. **Rectifier test.** Unplug the regulator/rectifier and check the forward and reverse bias of the diodes in the rectifier. Note that you should have two checks for testing a half-wave system, four checks for a full-wave system, and six checks for a three-phase system.

   a. forward bias: _____

   _____

   _____

   b. reverse bias: _____

   _____

   _____

9. Is this charging system working correctly?

   Yes   No   *Circle one*

10. If you circled "No," explain what is wrong with the system.

   _____

   _____

   _____

INSTRUCTOR VERIFICATION: _____

Job Sheet 14-3

Name: _____ Date: _____ Instructor: _____

Charging System Inspection #3

Objective

After completing this job sheet, you should be able to accurately evaluate any charging system.

Directions

Check at least one example of each type of charging system available. Since some tests listed on the job sheet will not apply to all charging systems, mark "NA" on tests that do not apply. Use your multimeter, a training aid designated for this job sheet, and the appropriate service manual to answer the questions below.

Tools Needed

Multi-meter

1. Identify the type of charging system by the number of leads on the stator.

   a. **1**, half wave

   b. **2**, full wave   *Circle one*

   c. **3**, three phase

2. Is this a **permanent-magnet** or an **excited-field** charging system? *Circle one*

3. **Key-off amp draw test.** Determine if the electrical system is draining the battery when the key is switched off by hooking your meter in series to the negative side of the battery and observing any current draw with the key off. Note that some systems with clocks or computer memory should draw 0.1–0.2 milliamps with the key off.

   Current draw: _____

4. **Charging voltage test.** Check the charging voltage at the battery. Hook your meter in parallel to the battery and let the vehicle idle.

   Voltage at idle: _____

5. **Charging current test.** Remove the main fuse and hook your meter in series at the fuse holder after you have configured the meter to test DCA. You should now be able to start the machine with the meter hooked up in series without damage to it. Observe the charging current at idle.

   Current at idle: _____

6. **Stator output test.** Unplug all the stator leads and hook the meter to one pair with the meter set at ACV. Test all three pairs of leads. Start the machine and record the voltage at idle.

   ACV at idle: 1–2 _____ ACV

   2–3 _____ ACV

   1–3 _____ ACV

7. **Field coil test.** If the system you are testing has an excited-field rotor, you may be able to check the field coil and brushes. Refer to the manual for details on testing the field coil.

   a. What is the manual spec for brush length? _____ mm

   b. What is the manual spec for field coil resistance? _____ ohms

8. **Rectifier test.** Unplug the regulator/rectifier and check the forward and reverse bias of the diodes in the rectifier. Note that you should have two checks for testing a half-wave system, four checks for a full-wave system, and six checks for a three-phase system.

   a. forward bias: _____

   _____

   _____

   b. reverse bias: _____

   _____

   _____

9. Is this charging system working correctly?

    Yes   **No**   *Circle one*

10. If you circled "No," explain what is wrong with the system.

   _____

   _____

   _____

INSTRUCTOR VERIFICATION: _____

Job Sheet 14-4

Name: _____ Date: _____ Instructor: _____

Charging System Inspection #4

Objective

After completing this job sheet, you should be able to accurately evaluate any power equipment charging system.

Directions

Check at least one example of each type of charging system available. Since some tests listed on the job sheet will not apply to all charging systems, mark "NA" on tests that do not apply. Use your multi-meter, a training aid designated for this job sheet, and the appropriate service manual to answer the questions below.

Tools Needed

Multi-meter

1. Identify the type of charging system by the number of leads on the stator.

    a. **1**, half wave

    b. **2**, full wave   *Circle one*

    c. **3**, three phase

2. Is this a **permanent-magnet** or **excited-field** charging system? *Circle one*

3. **Key-off amp draw test.** Determine if the electrical system is draining the battery when the key is switched off by hooking your meter in series to the negative side of the battery and observing any current draw with the key off. Note that some systems with clocks or computer memory should draw 1–2 milliamps with the key off.

    Current draw: _____

4. **Charging voltage test.** Check the charging voltage at the battery. Hook your meter in parallel to the battery and let the vehicle idle.

    Voltage at idle: _____

5. **Charging current test.** Remove the main fuse and hook your meter in series at the fuse holder after you have configured the meter to test DCA. You should now be able to start the machine with the meter hooked up in series without damage to it. Observe the charging current at idle.

    Current at idle: _____

6. **Stator output test.** Unplug all the stator leads and hook the meter to one pair with the meter set at ACV. Test all three pairs of leads. Start the machine and record the voltage at idle.

   ACV at idle: 1–2 _____ ACV

   2–3 _____ ACV

   1–3 _____ ACV

7. **Field coil test.** If the system you are testing has an excited-field rotor, you may be able to check the field coil and brushes. Refer to the manual for details on testing the field coil.
   a. What is the manual spec for brush length? _____ mm
   b. What is the manual spec for field coil resistance? _____ ohms

8. **Rectifier test.** Unplug the regulator/rectifier and check the forward and reverse bias of the diodes in the rectifier. Note that you should have two checks for testing a half-wave system, four checks for a full-wave system, and six checks for a three-phase system.

   a. forward bias: _____

   _____

   _____

   b. reverse bias: _____

   _____

   _____

9. Is this charging system working correctly?

   Yes   **No**   *Circle one*

10. If you circled "No," explain what is wrong with the system.

    _____

    _____

    _____

INSTRUCTOR VERIFICATION: _____

CHAPTER 14

Test

True/False

Indicate whether the statement is true or false.

_____ 1. A rectifier changes DC to AC using diodes.

_____ 2. An alternator has two basic parts, a rotor and a rectifier.

_____ 3. Excited-field alternators use brushes to energize the field coil.

_____ 4. Conventional and maintenance-free batteries vent through a small opening on the side.

_____ 5. A half-wave charging system uses a grounded charging coil.

Multiple Choice

Identify the choice that best completes the statement or answers the question.

_____ 6. A(n) _____ will display the service condition of a battery.
  a. conductance tester   c. voltmeter
  b. hydrometer   d. ammeter

_____ 7. _____ send electrical power to the field coil in an excited-field alternator.
  a. Carbon brushes   c. Permanent magnets
  b. Diodes   d. Rectifiers

_____ 8.   A _____ stator has three charging coils.

      a. half-wave   c. three-phase

      b. full-wave   d. DC

_____ 9.   The rectifier in a three-phase charging system contains _____ diodes.

      a. 2   c. 6

      b. 4   d. 8

_____ 10. The rotor in a three-phase excited-field charging system should have about _____ ohms of resistance when checked with an ohmmeter.

      a. 2   c. 6

      b. 4   d. 8

Short Answers

11. Explain why some power equipment systems have a charging system and describe how a basic charging system works. Your answer should address rectification and voltage regulation.

12. Describe how an alternator generates AC power.

13. Describe the difference between a permanent-magnet charging system and an excited-field charging system.

14. How are permanent and excited-field charging systems regulated?

15. What are the differences among a half-wave, full-wave, and three-phase stator coils and rectifiers?

INSTRUCTOR VERIFICATION: _____

CHAPTER 15

Ignition and Electric Starting Systems

Shop Assignment 15-1

Name: _____ Date: _____ Instructor: _____

Ignition Systems Knowledge Assessment

Objective
By correctly completing this assessment, you should be able to demonstrate your knowledge of charging system fundamentals.

Directions
Answer the questions below using your textbook as a resource.

1. Name the three functions of an ignition system as stated in the textbook.

   a. Provide a hot spark, prevent the spark plug tip from burning, provide a properly timed spark

   b. Maintain a spark long enough to ignite the fuel mixture, deliver a spark to each cylinder at the correct time, deliver a hot spark

   c. Maintain a spark long enough to ignite the fuel mixture, deliver a spark well in advance of top-dead center (TDC), deliver a spark that has ample reserve energy

2. What type of ignition system uses an AC power source?

   a. A points-type ignition system

   b. A capacitor discharge ignition system

   c. A battery-powered ignition system

   d. A digital ignition system

3. What type of ignition system fires the spark plug every 360° of crankshaft rotation?

   a. Wasted spark system

   b. Multiple discharge system

   c. Digital system

   d. A Hall effect system

4. A Hall effect sensor is used to _____ the spark.

   a. time

   b. increase

   c. decrease

   d. delay

5. Power equipment engines generally have a(n) _____ ignition system.

   a. battery-powered

   b. low-tension magneto

   c. high-tension magneto

   d. AC-powered

6. An ignition coil is essentially a _____.

   a. flux capacitor

   b. transformer

   c. zener diode

   d. little battery

7. A coil with more secondary windings than primary windings should

   a. step up the primary voltage.

   b. step down the primary voltage.

   c. step up the primary current.

   d. provide earlier ignition timing.

8. A trigger device works by switching the primary current to the ignition coil

   a. off and on.

   b. on as soon as the piston reaches TDC.

   c. off when the piston reaches TDC.

   d. to AC when the reluctor passes TDC.

9. A spark plug with a long heat path is _____ than a plug with a short heat path.

   a. hotter

   b. colder

   c. bigger

   d. smaller

10. Spark plugs have resistors built into them to

   a. make them last longer.

   b. increase secondary voltage.

   c. reduce radio frequency interference.

   d. reduce preignition.

11. Rare earth spark plugs are used to

   a. reduce the voltage needed to spark them and to last longer.

   b. increase the spark volume.

   c. keep the plug running cooler.

   d. keep the plug clean.

12. Breaker point ignition systems are no longer used because they

   a. are too complicated.

   b. require frequent maintenance.

   c. operate at too high a temperature.

   d. are too slow to deliver enough sparks to the plug.

13. When you see a set of points arcing, it means the

   a. coil is bad.

   b. points are mistimed.

   c. condenser is bad.

   d. spark plug resistor is bad.

14. The ignition switch on an AC-powered ignition system

   a. grounds the system to stop the engine.

   b. cuts the voltage to the primary side of the coil to stop the engine.

15. A starter motor can draw over 120 amps of current while cranking the engine over.

   a. True

   b. False

INSTRUCTOR VERIFICATION: _____

9.   A spark plug with a long heat path is _____ than a plug with a short heat path.

a.   hotter.

b.   colder.

c.   bigger.

d.   smaller.

10.   Spark plugs have resistors built into them to

a.   make them last longer.

b.   increase secondary voltage.

c.   reduce radio frequency interference.

d.   reduce preignition.

11.   Rare earth spark plugs are used to

a.   reduce the voltage needed to spark them and to last longer.

b.   increase the spark volume.

c.   keep the plug running cooler.

d.   keep the plug clean.

12.   Breaker point ignition systems are no longer used because they

a.   are too complicated.

b.   require frequent maintenance.

c.   operate at too high a temperature.

d.   are too slow to deliver enough sparks to the plug

13.   When you see a set of points arcing, it means the

a.   coil is bad.

b.   points are mistimed.

c.   condenser is bad.

d.   spark plug resistor is bad.

14.   The ignition switch on an AC-powered ignition system

a.   grounds the system to stop the engine.

b.   cuts the voltage to the primary side of the coil to stop the engine

15.   A starter motor can draw over 120 amps of current while cranking the engine over.

a.   True

b.   False

INSTRUCTOR VERIFICATION:

Job Sheet 15-1

Name: _____ Date: _____ Instructor: _____

AC CDI Ignition System Inspection

Objective

After completing this job sheet, you should be able to accurately evaluate an AC CDI ignition system.

Directions

Use your multi-meter, a peak voltage adapter, and a training aid designated for this lab sheet as well as the appropriate service manual to answer the questions below.

Tools Needed

Multi-meter with peak voltage adapter (Note: a peak voltage adapter is a simple device that allows your meter to accurately display the peak of any AC voltage you are measuring. You can use your multi-meter without one but the voltage displayed will be much lower than the actual voltage.)

1. Remove the spark plug and insert it into the plug cap. Lay the plug on the cylinder head and crank/pull the engine over with all switches in the "RUN" position. Did you see a spark?

   Yes   **No**   *Circle the answer*

2. Does the ignition and/or stop switch ground this system or cut battery voltage to the primary side of the ignition coil?

   a. Grounds the system

   b. Cuts current to the coil   *Circle the answer*

Exciter Coil Check

3. Disconnect the leads from the source or exciter coil and attach them to the peak voltage adapter. Attach the leads of your peak voltage adapter to your meter in the DCV jack and the common jack. Set the meter to "DCV" and turn the engine over briskly. Note that the exciter coil leads are found at the CDI module.

   Record the voltage: _____ ACV

   Factory spec for exciter coil voltage: _____ ACV

4. Set your meter to read ohms and measure the resistance of the exciter coil.

   Record the resistance: _____ ohm(s)

   Factory spec for exciter coil resistance: _____ ohm(s)

5. Reconnect the exciter coil leads.

Trigger Coil Check

6. Disconnect the leads from the pulser/trigger coil. Turn the engine over briskly. Note that the leads are usually found at the CDI module.

   Record the voltage: _____ ACV

   Factory spec for pulser coil voltage: _____ ACV

7. Set your meter to read ohms and measure the resistance of the trigger coil.

   Record the resistance: _____ ohms

   Factory spec for trigger coil resistance: _____ ohms

8. Reconnect the trigger coil leads.

Stop Switch Check

9. Disconnect the leads from the stop switch. Measure the resistance of the stop switch in both the On and Off positions.

   On: _____ ohm(s)

   Off: _____ ohm(s)

10. Is the stop switch wired in **series** or **parallel** on this vehicle? *Circle the answer*

11. Reconnect all leads and check to see if there is spark at the plug as you did in Step 1. Did you see a spark?

    Yes   **No**   *Circle the answer*

12. Did you find any components that did not meet factory spec?

    Yes   **No**   *Circle the answer*

13. Make the vehicle ready for final inspection.

INSTRUCTOR VERIFICATION: _____

Job Sheet 15-2

Name: _____ Date: _____ Instructor: _____

Magneto Ignition System Inspection

Objective

After completing this job sheet, you should be able to accurately evaluate the ignition system in an engine equipped with a magneto ignition system.

Directions

Use your multi-meter, a training aid designated for this lab sheet, and the appropriate service manual to answer the questions below.

Tools Needed

Multi-meter

1. Remove the spark plug and insert it into the plug cap. Lay the plug on the cylinder head and crank/kick the engine over with all switches in the "Run" position. Did you see a spark?

   Yes   **No**   *Circle the answer*

2. Does the ignition and/or stop switch ground this system at the primary side of the ignition coil?

   a. Grounds the system

   b. Cuts current to the coil   *Circle the answer*

Primary Voltage Check

3. Consult the schematic in the service manual to locate the source coil that supplies voltage to the ignition coil. Set your meter to ACV (DCV if using a peak voltage adapter) and check to see how many volts are present when turning the engine over.

   Record the voltage: _____ ACV

Points Check

4. Disconnect the breaker point leads and attach your ohmmeter to the points and the other lead to ground. Open and close the points manually and note the resistance reading.

   Open: _____ ohm(s)

   Closed: _____ ohm(s)

Stop Switch Check

5. Disconnect the leads from the stop switch. Measure the resistance of the stop switch in both the On and Off positions.

    On: _____ ohm(s)

    Off: _____ ohm(s)

6. Is the stop switch wired in **series** or **parallel** on this vehicle?   *Circle the answer*

7. Reconnect all leads and check to see if there is a spark at the plug as you did in Step 1. Did you see a spark?

    Yes   No   *Circle the answer*

8. Did you find any components that did not meet factory spec?

    Yes   No   *Circle the answer*

9. Make the vehicle ready for final inspection.

INSTRUCTOR VERIFICATION: _____

Job Sheet 15-3

Name: _____ Date: _____ Instructor: _____

Pointless Electronic Ignition System Inspection

Objective

After completing this job sheet, you should be able to accurately evaluate the ignition system on an electronic pointless ignition system.

Directions

Use your multi-meter, a training aid designated for this lab sheet, and the appropriate service manual to answer the questions below.

Tools Needed

Multi-meter

1. Remove the spark plug and insert it into the plug cap. Lay the plug on the cylinder head and crank/kick the engine over with all switches in the Run position. Did you see a spark?

   Yes   **No**   *Circle the answer*

2. Does the ignition and/or stop switch ground this system or cut voltage to the primary side of the ignition coil?

   a. Grounds the system

   b. Cuts current to the coil   *Circle the answer*

Primary Voltage Check

3. Remove the leads from the primary side of the ignition coil and check for primary voltage.

   Record the voltage: _____ DCV

4. Consult the service manual schematic to find what color leads supply 12 DCV to the ignition module. Remove those leads and check for battery voltage there.

   Record the voltage: _____ DCV

Trigger Coil Check

5. Disconnect the leads from the pulser/trigger coil. Turn the engine over briskly. Note that the leads are usually found at the ignition module.

   Record the voltage: _____ ACV

   Factory spec for pulser coil voltage: _____ ACV

6. Set your meter to read ohms and measure the resistance of the trigger coil.

   Record the resistance: _____ ohm(s)

   Factory spec for trigger coil resistance: _____ ohm(s)

7. Reconnect the trigger coil leads.

Stop Switch Check

8. Disconnect the leads from the stop switch. Measure the resistance of the stop switch in both the On and Off positions.

   On: _____ ohm(s)

   Off: _____ ohm(s)

9. Is the stop switch wired in **series** or **parallel** on this vehicle?   *Circle the answer*

10. Reconnect all leads and check to see if there is spark at the plug as you did in Step 1. Did you see a spark?

    Yes   **No**   *Circle the answer*

11. Did you find any components that did not meet factory spec?

    Yes   **No**   *Circle the answer*

12. Make the vehicle ready for final inspection.

INSTRUCTOR VERIFICATION: _____

CHAPTER

15

Test

True/False

Indicate whether the statement is true or false.

_____ 1. A wasted spark ignition system fires the spark plug once per revolution.

_____ 2. Higher octane gas is more easily ignited by the ignition system than lower octane fuel.

_____ 3. High turbulence in the combustion chamber requires that ignition be slightly retarded.

_____ 4.   When the engine is under a heavy load, the ignition timing must be retarded.

_____ 5. Cylinder number one always starts the firing order.

Multiple Choice

Identify the choice that best completes the statement or answers the question.

_____ 6. The ignition timing marks are located on the _____.
    a. flywheel   c. counter sprocket
    b. camshaft   d. connecting rod

_____ 7. Older ignition systems were equipped with a(n) _____ advance mechanism.
    a. electronic   c. centrifugal
    b. vacuum   d. manual

_____ 8. Power equipment ignition systems have just two power sources, the battery or a(n) _____ .

    a. capacitor

    b. AC generator

    c. pulser coil

    d. magnet

_____ 9. The first set of windings in an ignition coil is called the _____ winding.

    a. primary

    b. secondary

    c. induction

    d. field

_____ 10. Electronic trigger devices like Hall effect sensors completely eliminate the need for _____ .

    a. ignition coils

    b. pulse generators

    c. breaker points

    d. reluctors

Short Answers

11. What are the three functions of the ignition system?

12. How do higher rpm and increased turbulence in the combustion chamber affect ignition advance?

13. What are two advantages of an AC-powered ignition system over a battery-powered ignition system?

14. What advantage does a battery-powered ignition system have over an AC-powered ignition system?

15. Explain how the heat range of a spark plug is determined.

INSTRUCTOR VERIFICATION:

CHAPTER 16

Power Equipment Engine Maintenance

Shop Assignment 16-1

Name: _____ Date: _____ Instructor: _____

Maintenance Knowledge Assessment

Objective
By correctly completing this assessment, you should be able to demonstrate your knowledge of power equipment maintenance procedures.

Directions
Answer the questions below using your textbook as a resource.

1. A carburetor clean is usually part of a tune up.

   a. True

   b. False

2. You should check the specific gravity of a(n) _____ during a tune up.

   a. MF battery

   b. AGM battery

   c. conventional

   d. NiCad battery

3. The best source for the maintenance schedule for a piece of power equipment is found in the _____.

   a. manufacturer's Internet site

   b. owner's manual

   c. service manual

   d. b and c

4. It is not necessary to replace the oil filter with every oil change if the oil change interval is within the maintenance schedule.

   a. True

   b. False

5. Checking the oil level with a sight glass is done _____.

   a. only when the machine is upright and level

   b. upright or on an incline

   c. before the engine is started

6. A compression test cannot be done on a two-stroke engine.

   a. True

   b. False

7. Checking the spark plugs during a tune-up can tell you a great deal about the engine's condition.

   a. True

   b. False

8. The paper air filter on a high hour machine should be _____ during the course of a major tune up.

   a. blown out with compressed air

   b. washed

   c. replaced

   d. none of the above

INSTRUCTOR VERIFICATION:

Job Sheet 16-1

Name: _____ Date: _____ Instructor: _____

Four-Stroke Tune and Service

Objective

After completing this job sheet, you should be able to correctly service a power equipment engine according to the maintenance schedule and be able to perform a leak-down and compression test.

Directions

Using a training aid designated for this job sheet and the factory service manual, perform a normal service.

Tools Needed

Various hand tools, compression gauge, leak-down tester

1. Consult Table 16-1 on p. 336 of the textbook. Inspect the items listed in the maintenance schedule. If you have any questions concerning a service procedure, consult your instructor.

2. **Compression test**

    Manufacturer's specification: _____

    Measured compression: Cyl #1 _____

                                 Cyl #2 _____

3. **Leak-down test**

    Measured leak-down: Cyl #1 _____

                                 Cyl #2 _____

4. **Valve Clearance Inspection**

    Manufacturer's specification: Intake _____

                                    Exhaust _____

    Measured valve lash: Intake   Cyl #1 _____

                                    Cyl #2 _____

    Measured valve lash: Exhaust   Cyl #1 _____

                                    Cyl #2 _____

Note: Inspect, do not replace items listed below, and then check off the box signifying you have completed the inspection.

- Fuel line and tank ☐
- Throttle operation ☐
- Choke operation ☐
- Air cleaner ☐
- Spark plugs ☐
- Engine oil ☐
- Oil filter ☐
- Ignition timing ☐
- Idle speed ☐
- Coolant ☐
- Headlight ☐
- Blade brake clutch ☐
- Valve clearance ☐
- Spark arrestor ☐

5. Make the vehicle ready for final inspection.

Note any additional items that need service:

_____

INSTRUCTOR VERIFICATION:

Job Sheet 16-2

Name: _____ Date: _____ Instructor: _____

Two-Stroke Tune Up

Objective

After completing this job sheet, you should be able to correctly service a two-stroke power equipment engine according to the maintenance schedule and be able to perform a compression test.

Directions

Using a training aid designated for this job sheet and the factory service manual, perform a normal service.

Tools Needed

Various hand tools, compression gauge

1. Consult Table 16-1 on p. 336 of the textbook. Inspect the items listed in the maintenance schedule. If you have any questions concerning a service procedure, consult your instructor.

2. **Compression test**

   Manufacturer's specification: _____

   Measured compression: Cyl #1 _____

   Cyl #2 _____

3. Note: Inspect, do not replace items listed below, and then check off the box signifying you have completed the inspection.

   - Fuel line and tank   ☐
   - Throttle operation   ☐
   - Choke operation   ☐
   - Air cleaner   ☐
   - Spark plugs   ☐
   - Engine oil   ☐
   - Oil filter   ☐
   - Carb synch   ☐
   - Idle speed   ☐
   - Coolant   ☐
   - Headlight   ☐
   - Blade brake clutch   ☐
   - Valve clearance   ☐
   - Spark arrestor   ☐
   - Injector tank   ☐

4. Make the unit ready for final inspection.

   Note any additional items that need service:
   _____

INSTRUCTOR VERIFICATION: _____

CHAPTER 16

Test

True/False

Indicate whether the statement is true or false.

_____ 1. A tune up usually includes a carburetor overhaul.

_____ 2. Always clean the unit before performing scheduled maintenance.

_____ 3. The compression test by itself is seldom a good reason to disassemble the engine.

_____ 4. Oil changes are among the most frequent services done at any power equipment repair facility.

_____ 5. Some small engines do not have a drain plug.

Multiple Choice

Identify the choice that best completes the statement or answers the question.

_____ 6. Coolant around the _____ signals that the water pump seal is failing.

    a. reservoir cap   c. telltale hole

    b. radiator cap   d. thermostat

_____ 7. Most manufacturers recommend a coolant flush every _____ months.

    a. 2   c. 24

    b. 12   d. 36

_____ 8. Recommended compression readings can be found in the _____.

    a. owner's manual   c. service manual

    b. manufacturer's Web site   d. info tag on the frame

_____ 9. _____ is not a valve adjustment method.

    a. Screw and lock nut   c. Hydraulic

    b. Tip grinding   d. Lever and pulpit

_____ 10. Oil-fouled spark plugs are an indication that the _____ may be plugged up.

    a. crankcase breather   c. purge control valve

    b. exhaust port   d. fuel injectors

INSTRUCTOR VERIFICATION:

Power Equipment Engine Troubleshooting

CHAPTER 17

Shop Assignment 17-1

Name: _____ Date: _____ Instructor: _____

Troubleshooting Knowledge Assessment

Objective
By correctly completing this assessment, you will demonstrate your knowledge of the principles of effective troubleshooting.

Directions
Answer the questions below using your textbook as a resource.

1. Name the three categories of failures.

    1. _____

    2. _____

    3. _____

2. A customer comes into your shop with a constant failure. You have verified the failure. What is the next step?

    a. Repair the problem.

    b. Consult the service manual troubleshooting flow chart.

    c. Call the help line.

    d. Isolate the problem.

3. The service manager gives you a repair order that says the machine won't start. Where should you begin in troubleshooting the problem?

    a. Do a complete ignition system diagnosis.

    b. Remove and clean the carburetors.

    c. Check the engine stop switch and the fuel tank.

    d. Check the compression and leak down.

4. A customer brings his HT3813 lawn tractor into a Honda service department with a dead battery. The technician has seen this problem before. A continuity tests shows the stator is bad, so he replaces the stator, charges the battery, flags the ticket, and calls the customer to pick it up. However, the customer calls back the next day and says his battery is dead again. What basic troubleshooting step did the technician neglect to perform?

    a. He didn't replace the battery.

    b. He didn't do a test ride.

    c. He didn't verify the repair.

    d. Replace the rectifier

5. A customer comes into your shop with a machine that runs poorly at high engine speed. She tells you that he has had this vehicle at three shops previously and they all cleaned and rejetted the carbs. Where would you start your troubleshooting procedure?

    a. Remove the carbs to see if they really are clean inside.

    b. Remove the carbs to check the jets.

    c. Call the other shops to see if they have any ideas.

    d. Assume the carbs are clean and the jetting correct and begin isolating the problem.

6. A technician is working on an ignition system with no spark. He replaces the plugs, battery, and the ignition coil but still no spark is found. He then pulls a new mower off the showroom floor, removes its ignition system, and puts it on the engine he is working on to no avail. What mistake did he make in troubleshooting?

    a. He changed too many things at once.

    b. He assumed the ignition system was at fault without properly isolating the problem.

    c. He didn't think out the possibilities before doing something drastic.

    d. All the above.

7. The most difficult problems to troubleshoot are _____.

    a. electrical

    b. shifting

    c. fuel related

    d. handling related

8. Most electrical problems are related to the _____.

   a. wiring connections

   b. charging system

   c. ignition system

   d. fuel injection system

9. You suspect that a "no spark" condition is caused by the engine stop switch. How should you verify your suspicion?

   a. Replace it with a known good switch.

   b. Order a new switch and replace it.

   c. Check it with an ohmmeter.

   d. Cycle the switch off and on repeatedly.

10. The most difficult electrical problems to troubleshoot are _____.

    a. ignition system problems

    b. intermittent problems

    c. fuel injection problems

    d. constant problems

INSTRUCTOR VERIFICATION:

8. Most electrical problems are related to the _____

   a. wiring connections

   b. charging system

   c. ignition system

   d. fuel injection system

9. You suspect that a "no spark" condition is caused by the engine stop switch. How should you verify your suspicion?

   a. Replace it with a known good switch.

   b. Order a new switch and replace it.

   c. Check it with an ohmmeter.

   d. Cycle the switch off and on repeatedly.

10. The most common electrical problems to troubleshoot are _____

    a. ignition system problems

    b. intermittent problems

    c. fuel injection problems

    d. constant problems

INSTRUCTOR VERIFICATION. _____

Job Sheet 17-1

Name: _____ Date: _____ Instructor: _____

Troubleshoot "No Start"

Objective

After completing this job sheet in less than 1 hour, you will have demonstrated your mastery of the troubleshooting procedures necessary to diagnose the customer's concern.

Tools Needed

Various hand tools

Customer concern: The engine will not start.

1. Write down the troubleshooting steps used during your inspection of this problem.

   _____

   _____

   _____

2. How would you correct the problem?

   _____

INSTRUCTOR VERIFICATION: _____

Job Sheet 17-1

Name: _____   Date: _____   Instructor: _____

Troubleshoot "No Start"

Objective

After completing this job sheet in less than 1 hour, you will have demonstrated your mastery of the troubleshooting procedures necessary to diagnose the customer's concern.

Tools Needed

Various hand tools

Customer concern: The engine will not start.

1. Write down the troubleshooting steps used during your inspection of this problem.

_____

_____

_____

2. How would you correct the problem?

_____

INSTRUCTOR VERIFICATION _____

Job Sheet 17-2

Name: _____ Date: _____ Instructor: _____

Troubleshoot "Engine Won't Turn Over"

Objective

After completing this job sheet in less than 1 hour, you will have demonstrated your mastery of the troubleshooting procedures necessary to diagnose the customer's concern.

Tools Needed

Various hand tools, multi-meter

Customer concern: An electric starter will not turn over. The battery has just been replaced.

1. Write down the troubleshooting steps used during your inspection of this problem.

   _____

   _____

   _____

2. How would you correct the problem?

   _____

INSTRUCTOR VERIFICATION: _____

Job Sheet 17-2

Troubleshoot "Engine Won't Turn Over"

Name: _____ Date: _____ Instructor: _____

Objective

After completing this job sheet in less than 1 hour, you will have demonstrated your mastery of the troubleshooting procedures necessary to diagnose the customer's concern.

Tools Needed

Various hand tools, multi-meter

Customer concern: An electric starter will not turn over. The battery has just been replaced.

1. Write down the troubleshooting steps used during your inspection of this problem.

_____

_____

_____

2. How would you correct the problem?

_____

INSTRUCTOR VERIFICATION

Job Sheet 17-3

Name: _____ Date: _____ Instructor: _____

Troubleshoot "Engine Overheats"

Objective

After completing this job sheet in less than 1 hour, you will have demonstrated your mastery of the troubleshooting procedures necessary to diagnose the customer's concern.

Tools Needed

Various hand tools, multi-meter

Customer concern: Engine overheats.

1. Write down the troubleshooting steps used during your inspection of this problem.

   _____

   _____

   _____

2. How would you correct the problem?

   _____

INSTRUCTOR VERIFICATION: _____

Name: _____ Date: _____ Instructor: _____

Troubleshoot "Engine Overheats."

Objective

After completing this job sheet in less than 1 hour, you will have demonstrated your mastery of the troubleshooting procedures necessary to diagnose the customer's concern.

Tools Needed

Various hand tools, multi-meter.

Customer concern: Engine overheats.

1. Write down the troubleshooting steps used during your inspection of this problem:

_____

_____

2. How would you correct the problem?

_____

INSTRUCTOR VERIFICATION: _____

Job Sheet 17-4

Name: _____ Date: _____ Instructor: _____

Troubleshoot "Smokes Excessively"

Objective
After completing this job sheet in less than 1 hour, you will have demonstrated your mastery of the troubleshooting procedures necessary to diagnose the customer's concern.

Tools Needed
Various hand tools

Customer concern: Engine smokes.

1. Write down the troubleshooting steps used during your inspection of this problem.

   _____

   _____

   _____

2. How would you correct the problem?

   _____

INSTRUCTOR VERIFICATION: _____

Name: _____ Date: _____ Instructor: _____

Troubleshoot "Smokes Excessively"

Objective

After completing this job sheet in less than 1 hour, you will have demonstrated your mastery of the troubleshooting procedures necessary to diagnose the customer's concern.

Tools Needed

Various hand tools

Customer concern: Engine smokes.

1. Write down the troubleshooting steps used during your inspection of this problem.

_____

_____

_____

2. How would you correct the problem?

INSTRUCTOR VERIFICATION

Job Sheet 17-5

Name: _____ Date: _____ Instructor: _____

Troubleshoot "Engine Won't Idle"

Objective

After completing this job sheet in less than 1 hour, you will have demonstrated your mastery of the troubleshooting procedures necessary to diagnose the customer's concern.

Tools Needed

Various hand tools

Customer concern: Engine won't idle after sitting all winter.

1. Write down the troubleshooting steps used during your inspection of this problem.

_____

_____

_____

2. How would you correct the problem?

_____

INSTRUCTOR VERIFICATION: _____

Job Sheet 17-5

Name _____ Date: _____ Instructor: _____

Troubleshoot "Engine Won't Idle"

Objective

After completing this job sheet in less than 1 hour, you will have demonstrated your mastery of the troubleshooting procedures necessary to diagnose the customer's concern.

Tools Needed

Various hand tools

Customer concern: Engine won't idle after sitting all winter.

1. Write down the troubleshooting steps used during your inspection of this problem.

_____

_____

_____

2. How would you correct the problem?

_____

INSTRUCTOR VERIFICATION: _____

Job Sheet 17-6

Name: _____ Date: _____ Instructor: _____

Troubleshoot "Engine Surges at Idle"

Objective

After completing this job sheet in less than 1 hour, you will have demonstrated your mastery of the troubleshooting procedures necessary to diagnose the customer's concern.

Tools Needed

Various hand tools

Customer concern: Engine surges at idle.

1. Write down the troubleshooting steps used during your inspection of this problem.

    _____

    _____

    _____

2. How would you correct the problem?

    _____

INSTRUCTOR VERIFICATION: _____

Job Sheet 17-6

Name: _____ Date: _____ Instructor: _____

Troubleshoot "Engine Surges at Idle"

Objective

After completing this job sheet in less than 1 hour, you will have demonstrated your mastery of the troubleshooting procedures necessary to diagnose the customer's concern.

Tools Needed

Various hand tools

Customer concern: Engine surges at idle.

1. Write down the troubleshooting steps used during your inspection of this problem.

_____

_____

_____

2. How would you correct the problem?

_____

INSTRUCTOR VERIFICATION

Job Sheet 17-7

Name: _____ Date: _____ Instructor: _____

Troubleshoot "FI Light On"

Objective
After completing this job sheet in less than 1 hour, you will have demonstrated your mastery of the troubleshooting procedures necessary to diagnose the customer's concern.

Tools Needed
Various hand tools

Customer concern: FI light on.

1. Write down the troubleshooting steps used during your inspection of this problem.

   _____

   _____

   _____

2. How would you correct the problem?

   _____

INSTRUCTOR VERIFICATION: _____

Name: _____   Date: _____   Instructor: _____

Troubleshoot "FI Light On"

Objective

After completing this job sheet in less than 1 hour, you will have demonstrated your mastery of the motorcycle troubleshooting procedures necessary to diagnose the customer's concern.

Tools Needed

Various hand tools

Customer concern: FI light on.

1. Write down the troubleshooting steps used during your inspection of this problem.

_____

_____

_____

2. How would you correct the problem?

_____

Job Sheet 17-8

Name: _____ Date: _____ Instructor: _____

Troubleshoot "FI Light Came On, Now Off"

Objective

After completing this job sheet in less than 1 hour, you will have demonstrated your mastery of the troubleshooting procedures necessary to diagnose the customer's concern.

Tools Needed

Various hand tools

Customer concern: FI light came on but is now off.

1. Write down the troubleshooting steps used during your inspection of this problem.

   _____

   _____

   _____

2. How would you correct the problem?

   _____

INSTRUCTOR VERIFICATION: _____

Job Sheet 17-8

Troubleshoot "FI Light Came On, Now Off"

Name: _____ Date: _____ Instructor: _____

Objective

After completing this job sheet in less than 1 hour, you will have demonstrated your mastery of the troubleshooting procedures necessary to diagnose the customer's concern.

Tools Needed

Various hand tools

Customer concern: FI light came on but is now off.

1. Write down the troubleshooting steps used during your inspection of this problem.

_____

_____

_____

2. How would you correct the problem?

_____

INSTRUCTOR VERIFICATION: _____

Job Sheet 17-9

Name: _____ Date: _____ Instructor: _____

Troubleshoot "Battery Goes Dead"

Objective

After completing this job sheet in less than 1 hour, you will have demonstrated your mastery of the troubleshooting procedures necessary to diagnose the customer's concern.

Tools Needed

Various hand tools, multi-meter

Customer concern: Battery goes dead.

1. Write down the troubleshooting steps used during your inspection of this problem.

   _____

   _____

   _____

2. How would you correct the problem?

   _____

INSTRUCTOR VERIFICATION: _____

Job Sheet 17-9

Name: _____   Date: _____   Instructor: _____

Troubleshoot "Battery Goes Dead"

Objective

After completing this job sheet in less than 1 hour, you will have demonstrated your mastery of the troubleshooting procedures necessary to diagnose this customer's concern.

Tools Needed

Various hand tools, multi-meter

Customer concern: Battery goes dead.

1. Write down the troubleshooting steps used during your inspection of this problem.

_____

_____

_____

2. How would you correct the problem?

_____

INSTRUCTOR VERIFICATION: _____

Job Sheet 17-10

Name: _____ Date: _____ Instructor: _____

Troubleshoot "No Spark"

Objective
After completing this job sheet in less than 1 hour, you will have demonstrated your mastery of the troubleshooting procedures necessary to diagnose the customer's concern.

Tools Needed
Various hand tools, multi-meter

Customer concern: No spark.

1. Write down the troubleshooting steps used during your inspection of this problem.

   _____

   _____

   _____

2. How would you correct the problem?

   _____

INSTRUCTOR VERIFICATION: _____

Job Sheet 17-10

Name: _____ Date: _____ Instructor: _____

Troubleshoot "No Spark"

Objective

After completing this job sheet in less than 1 hour, you will have demonstrated your mastery of the troubleshooting procedures necessary to diagnose the customer's concern.

Tools Needed

Various hand tools, multi-meter

Customer concern: No spark.

1. Write down the troubleshooting steps used during your inspection of this problem.

_____

_____

_____

2. How would you correct the problem?

_____

INSTRUCTOR VERIFICATION _____

Job Sheet 17-11

Name: _____ Date: _____ Instructor: _____

Troubleshoot "Dim Headlight"

Objective

After completing this job sheet in less than 1 hour, you will have demonstrated your mastery of the troubleshooting procedures necessary to diagnose the customer's concern.

Tools Needed

Various hand tools, multi-meter

Customer concern: Dim headlight.

1. Write down the troubleshooting steps used during your inspection of this problem.

   _____

   _____

   _____

2. How would you correct the problem?

   _____

INSTRUCTOR VERIFICATION: _____

Job Sheet 17-11

Name: _____ Date: _____ Instructor: _____

Troubleshoot "Dim Headlight"

Objective
After completing this job sheet in less than 1 hour, you will have demonstrated your mastery of the troubleshooting procedures necessary to diagnose the customer's concern

Tools Needed
Various hand tools, multi-meter

Customer concern: Dim headlight.

1. Write down the troubleshooting steps used during your inspection of this problem.

_____

_____

_____

2. How would you correct the problem?

_____

INSTRUCTOR VERIFICATION: _____

Job Sheet 17-12

Name: _____ Date: _____ Instructor: _____

Troubleshoot "Abnormal Noise"

Objective

After completing this job sheet in less than 1 hour, you will have demonstrated your mastery of the troubleshooting procedures necessary to diagnose the customer's concern.

Tools Needed

Various hand tools

Customer concern: Abnormal noise.

1. Write down the troubleshooting steps used during your inspection of this problem.

   _____

   _____

   _____

2. How would you correct the problem?

   _____

   _____

   _____

INSTRUCTOR VERIFICATION: _____

Job Sheet 17-12

Name _____ Date _____ Instructor _____

Troubleshoot "Abnormal Noise"

Objective
After completing this job sheet in less than 1 hour, you will have demonstrated your mastery of the troubleshooting procedures necessary to diagnose the customer's concern.

Tools Needed
Various hand tools

Customer concern: Abnormal noise.

1. Write down the troubleshooting steps used during your inspection of this problem.

_____

_____

_____

2. How would you correct the problem?

_____

_____

_____

INSTRUCTOR VERIFICATION: _____